DICTIONARY OF

MILITARY ABBREVIATIONS

DICTIONARY OF

MILITARY ABBREVIATIONS

Norman Polmar

Mark Warren

Eric Wertheim

Naval Institute Press / Annapolis, Maryland

©1994
by Norman Polmar

All rights reserved. No part of this book may be reproduced without written permission.

Library of Congress Cataloging-in-Publication Data

Polmar, Norman.
    Dictionary of military abbreviations / Norman
  Polmar, Mark Warren, Eric Wertheim.
         p.   cm.
    ISBN 1-55750-680-9 (acid-free paper)
    1. Military art and science—Abbreviations. I. Warren,
  Mark, 1970–   . II. Wertheim, Eric, 1973–
  III. Title.
  U26.P615   1993
  355'.00148—dc20   93-34566
                                                                   CIP
Printed in the United States of America on acid-free paper ∞

9 8 7 6 5 4 3 2
First printing

Dedicated to Patricia Lee Lewis
Submariner par Excellence

The beginning of wisdom is calling things by their right names.
        Confucius

How many a dispute could have been deflated into a single paragraph if the disputants had just dared to define their terms.
        Aristotle

Gobbledygook: . . . the way the users of Latin phrases and big words, the double-talkers and the long-winded writers were moving in on us like an invisible empire.
        Maury Maverick
        Former U.S. representative from
        Texas and vice chairman,
        U.S. War Production Board,
        1944

Then, earlier this morning, the amphibious assault phase began with troops and equipment moving ashore by helicopter and the two types of amphibious landing craft, both what are called the LCACs and the LCUs. Now if only one knew what those stood for!
        Pentagon spokesman Pete Williams
        DOD news briefing, 4 August 1992

Contents

	Preface	xi
CHAPTER 1	**Military Abbreviations**	1
CHAPTER 2	**Aircraft Designations**	261
CHAPTER 3	**Aviation Unit Designations**	275
CHAPTER 4	**Military Ranks**	281
CHAPTER 5	**Missile and Rocket Designations**	291
CHAPTER 6	**Ship Designations**	297

Preface

The Goldwater-Nichols defense reorganization act passed by Congress in 1986 requires that U.S. military officers serve on a joint or unified staff as a prerequisite for promotion to flag or general officer. This dictionary is intended to assist those U.S. military officers who are suddenly "forced" to deal with the plethora of abbreviations used by the other services—as well as their own.

We have attempted to compile a dictionary of specific abbreviations that are currently in use, or that may not be current but will still be found in contemporary military documents. We have carefully avoided the "vacuum cleaner" or "garbage pail" approach to abbreviations, that of collecting every abbreviation that can be found. We have also made an effort to avoid obscure abbreviations, such as those relating to the subsystems of a specific weapon or sensor.

Many organizational abbreviations are not provided although their components should be found in chapter 1. For example, COMDESRON is not found in the book although its component abbreviations are:

COM	Commander
DES	Destroyer
RON	Squadron

Many organizations of all services are being realigned and some activities were being changed as this volume went to press.

PREFACE

As shown in the Contents, we have divided the abbreviations into several categories to help the user find specific sets of abbreviations. Services or agencies are identified for abbreviations that are unique to specific organizations. The services and agencies are:

DIA	Defense Intelligence Agency
DOD	Department Of Defense
JCS	Joint Chiefs of Staff
NASA	National Aeronautics and Space Administration
NATO	North Atlantic Treaty Organization
USA	U.S. Army
USAF	U.S. Air Force
USCG	U.S. Coast Guard
USMC	U.S. Marine Corps
USN	U.S. Navy

A note of observation: The use of capital letters, ampersands, dashes, and other punctuation marks and elements of style are often arbitrary in official abbreviations, in part because the abbreviations date from different eras (and hence different English usage) and were originated by different services and groups within those services. The abbreviations do not always highlight each word in the term. See, for example, DAWIA (Defense Aquisition Work Force Improvement Act). The word *force* is not included in the acronym.

The authors are grateful to several individuals who have taken the time to assist us in this project, especially Chief Warrant Officer Luke Arant, USA; Lt. Comdr. Kenneth Satterfield, USN, Department of Defense public affairs; and Peter Mersky of *Approach* magazine. Several members of the Naval Institute Staff provided invaluable support to us, especially Jaci Day, Eve Secunda, Cynthia Pierce, Mary Beth Straight, and Patty M. Maddocks. The perceptive editing of Marilyn Wilderson and Linda O'Doughda also contributed to the book. Tim Laur, Steve Llauso, and Patty Byars of the USNI Military Database helped with the initial compilation of this dictionary.

This book is, in part, a product of the Naval Institute's summer intern program. That effort could not have been undertaken without the early and strong support of Jim Sutton, former director of

PREFACE

marketing for the Naval Institute Press, Fred Rainbow, editor in chief of the Naval Institute *Proceedings*, and Barbara Broadhurst, Naval Institute personnel officer.

This book will be updated and reprinted periodically. Accordingly, additions or suggestions should be addressed to the authors at the Naval Institute Press, 118 Maryland Avenue, Annapolis, Maryland, 21402-5035.

Norman Polmar
Mark Warren
Eric Wertheim

DICTIONARY OF

MILITARY ABBREVIATIONS

CHAPTER I

Military Abbreviations

A

A1
    Staff Officer for Administration (USAF)
A2
    Staff Officer for Intelligence (USAF)
A^2C^2
    Army Airspace Command and Control (USA)
A3
    Staff Officer for Operations and Plans (USAF)
A4
    Staff Officer for Communications (USAF)
A-109
    OMB Circular A-109
A/A
    Angle of Attack
AA
    (1) Administrative Assistant
    (2) Air Assault (USA)

Note: * = numerical designation follows
      ** = normally used as word rather than abbreviation

AA-( )
    (3) Anti-Aircraft
    (4) Assembly Area (USA)
AA-( )*
    Designation for Soviet-Russian Air-to-Air missile (NATO)
AAA
    (1) Advanced Amphibious Assault (USMC)
    (2) Anti-Aircraft Artillery
    (3) Arrival and Assembly Area (USMC)
AA&A
    Armor, Armament, and Ammunition
AAAM
    Advanced Air-to-Air Missile [cancelled 1991]
AAAV
    Advanced Assault Amphibian Vehicle (USMC)
AAB
    (1) Aircraft Accident Board
    (2) Anti-Aircraft Battery
AABNCP
    Advanced Airborne National Command Post [see also ABNCP; E-4B aircraft]
AAC
    (1) Aerial Ambulance Company
    (2) Air [traffic] Area Control
    (3) Alaskan Air Command
AACC
    Area Approach Control Center
AACD
    Antenna Adjustable Current Distribution
AACE
    (1) Airborne Alternate Command Echelon
    (2) Air-to-Air Combat Environment
AADC
    Area Air Defense Commander
AADEOS
    Advanced Air Defense Electro-Optical Sensor (USA)
AADS
    Army Aviation Decontamination Station
AAE
    Army Acquisition Executive
AAFCE
    Allied Air Forces Central Europe (NATO)

AAFES
 Army and Air Force Exchange Service
AAFIF
 Automated Air Facilities Information File
AAG
 Aeromedical Airlift Group (USAF)
AAH
 Advanced Attack Helicopter
AAM
 Air-to-Air Missile
AAMP
 Army Aviation Modernization Plan
AAO
 Authorized Acquisition Objective
AAOE
 Arrival and Assembly Operation Element (USMC)
AAOG
 Arrival and Assembly Operation Group (USMC)
AAR
 (1) After Action Review
 (2) Aircraft Accident Report
AASP
 Arrival and Assembly Support Party (USMC)
AAV
 (1) Assault Amphibian Vehicle [formerly LVT] (USMC)
 (2) Autonomous Air Vehicle [drone—formerly RPV]
AAVS
 Aerospace Audiovisual Service
AAW
 Anti-Air Warfare
AAWC
 (1) Anti-Air Warfare Commander
 (2) Anti-Air Warfare Coordinator (USN)
AAWS-H
 Anti-Armor Weapon System (Heavy) (USA)
AAWS-M
 (1) Advanced Antitank Weapon System (Medium) (USA)
 (2) Anti-Armor Weapon System (Medium) (USA)
AB
 (1) Afterburner
 (2) Air Base
 (3) Aviation Boatswain's Mate

ABC
   Air Battle Captain (USA)
ABCCC
   Airborne Battlefield Command and Control Center [aircraft]
ABCP
   Airborne Command Post
ABDR
   Aircraft Battle Damage Repair (USAF)
ABF
   Advanced Bomb Family (USN)
ABFDS
   Aerial Bulk Fuel Delivery System (USAF)
ABG
   Air Base Group
ABL
   (1) Airborne Laser
   (2) Armored Box Launcher [Tomahawk]
ABM
   Anti-Ballistic Missile [properly used as adjective]
Abn
   Airborne
ABNCP
   Airborne National Command Post [see also AABNCP; E-4B aircraft] (USAF)
A-box
   fire support box; subdivision of a kill box (USMC)
ABS
   American Bureau of Shipping
AC
   (1) Active Component [of service]
   (2) Aircraft
   (3) Aircraft Commander
ACA
   (1) Airspace Control Authority (USAF)
   (2) Airspace Coordination Area (USA)
ACAA
   Automatic Chemical Agent Alarm
ACALS
   Army Computer-Aided Acquisition and Logistics Support
ACAP
   (1) Army Career and Alumni Program
   (2) Army Cost Analysis Paper

ACAT
- (1) Acquisition Category
- (2) Ashore Coordinated ASW Training (USN)

ACC
- (1) Air Combat Command [established 1992] (USAF)
- (2) Air Component Commander (USAF)
- (3) Airlift Coordination Center (USAF)
- (4) Airspace Coordination Center (NATO)
- (5) Arab Cooperation Council
- (6) Army Component Commander

ACCESS
Automated Command Control Executive Support System (USAF)

ACCHAN
Allied Command Channel [see CINCHAN]

ACCS
Army Command and Control Systems

ACDA
Arms Control and Disarmament Agency

ACDS
Advanced Combat Direction System

ACDUTRA
Active Duty for Training

ACE
- (1) Acquisition Enhancement Program
- (2) Airborne Command Element (USAF)
- (3) Air Combat Element (NATO)
- (4) Allied Command Europe (NATO)
- (6) Armored Combat Earthmover [M-9] (USA-USMC)
- (7) Armored Combat Equipment
- (8) Avenger Control Electronic (USA)
- (9) Aviation Combat Element (USMC)

ACEVAL
Air Combat Evaluation (USAF-USN)

ACF
Airlift Contingency Forces (USMC)

ACI
Allocated Configuration Identification

ACIB
Air Characteristic Improvement Board (USN)

ACINT
Acoustic Intelligence (USN)

ACIP
Aviation Career Incentive Pay (USN)
ACIPS
Accoustic Information Processing System (USN)
ACL
Allowable Cabin Load
ACLANT
Allied Command Atlantic (NATO)
ACLS
Automatic Carrier Landing System
ACM
(1) Advanced Cruise Missile
(2) Air Combat Maneuver
(3) Air Combat Maneuvering
ACMC
Assistant Commandant of the U.S. Marine Corps
ACMI
Air Combat Maneuvering Instrumentation [range]
ACMR
Air Combat Maneuvering Range
ACNO
Assistant Chief of Naval Operations (USN)
ACO
(1) Administrative Contracting Officer
(2) Air Control Officer (USN)
ACOE
Army Communities Of Excellence
ACOS
Assistant Chief Of Staff
ACP
(1) Aerial Communications Point (USA)
(2) Aerial Control Point (USA)
(3) Allied Communications Publications (NATO)
(4) Asset Capitalization Program (USAF)
(5) Aviation Continuation Pay (USN)
ACQ STRAT
Acquisition Strategy
ACR
Armored Cavalry Regiment (USA)
ACS
(1) African Coastal Security
(2) Army Community Service

(3) Artillery Computer System
(4) Assistant Chief of Staff
ACSC
Air Command and Staff College (USAF)
ACSFOR
Assistant Chief of Staff
(Force Development) (USA)
ACS/I
Assistance Chief of Staff for Intelligence (USAF)
ACSIM
Assistant Chief of Staff (Information Management)
ACSN
Advance Change Study Notice
ACT
Air Cavalry Troop (USA)
ACTI
Air Combat Tactics Instructor
ACTS
Advanced Contingency Theater Sensor (USA)
ACU
Assault Craft Unit (Navy)
ACV
(1) Air Cushion Vehicle
(2) Armored Combat Vehicle
ACW
Anti-Carrier Warfare (USN)
ACWP
Actual Cost of Work Performed
AD
(1) Advanced Deployability Posture
(2) Advanced Development
(3) Air Defense
(4) Armament Division (USAF)
Ada
computer programming language (DOD)
ADA
Air Defense Artillery
A/DACG
Arrival/Departure Airfield Control Group
ADAR
Air Deployable Active Reservoir

ADATS
　Air Defense Anti-Tank System (USA)
ADAWS
　Action Data Automated Weapon System [British]
ADCAP
　Advanced Capability [improved Mk 48 torpedo] (USN)
ADCOM
　Aerospace Defense Command
ADDISS
　Advanced Deployable Digital Imagery Support System
ADDS
　Army Data Distribution System
ADF
　(1) Air Defense Fighter
　(2) Australian Defence Force
　(3) Automatic Direction Finder (USN)
ADI
　Air Defense Initiative [part of SDI program]
ADIZ
　Air Defense Identification Zone
ADL
　Area Dental Laboratories (USA)
ADLAT
　Advanced Low Altitude Terrain-following [missile]
ADM
　(1) Acquisition Decision Memorandum
　(2) Advanced Development Model
　(3) Atomic Demolition Munition
ADNET
　Anti-Drug Network
ADOC
　Air Defense Operations Center
ADP
　Automatic Data Processing
ADPA
　American Defense Preparedness Association
ADPE
　ADP Equipment
ADR
　Aircraft Damage Repair (USN)
ADST
　Advanced Distributed Simulation Technology (USA)

ADU
 Auxiliary Display Unit
ADVCAP
 Advanced Capability [EA-6B Prowler aircraft modification] (USN)
AE
 (1) Acquisition Executive
 (2) Assault Echelon [of amphibious force] (USMC)
AEB
 Active Electronic Buoy [AN/SSQ-95(V)] (USN)
AEC
 Atomic Energy Commission (now Department of Energy)
AECA
 Arms Export Control Act
AECC
 Aeromedical Evacuation Control Center (USAF)
AED
 Aeronautical Engineering Division (USAF)
AEDC
 Arnold Engineering Development Center (USAF)
AEDO
 Aeronautical Engineering Duty Officer (USN)
AER
 Army Emergency Relief
AERS
 Army Education Requirement System
AETC
 Air Education and Training Command [established 1993] (USAF)
AEW
 Airborne Early Warning
AEW&C
 Airborne Early Warning and Control
AF
 Air Force [USAF preferred]
AFA
 Air Force Association
AFAA
 Air Force Audit Agency
AFAC
 Airborne Forward Air Control [or controller]
AFAE
 Air Force Acquisition Executive

AFAFC
Air Force Accounting and Finance Center
AFAL
Air Force Avionics Laboratory
AFALC
Air Force Air Logistics Center
AFAP
Artillery-Fired Atomic Projectiles
AFARV
Armored Forward Area Rearm Vehicle (USA)
AFAS
Advanced Field Artillery System
AFATDS
Advanced (or Army) Field Artillery Tactical Data System (USA)
AFATL
Air Force Armament Testing Laboratory
AFB
Air Force Base
AFBCMR
Air Force Board for Correction of Military Records
AFBDA
Air Force Base Disposal Agency
AFBS
Air Force Broadcasting Service
AFCAA
Air Force Cost Analysis Agency
AFCARA
Air Force Civilian Appellate Review Agency
AFCC
(1) Air Force Communications Command
(2) Air Force Component Commander
AF-CCP
Air Force Consolidation and Containerization Points
AFCEA
Armed Forces Communications and Electronics Association
AFCEE
Air Force Center for Environmental Excellence
AFCENT
Allied Forces Central Europe (NATO)
AFCESA
Air Force Civil Engineering Support Agency

AFCLC
   Air Force Contract Law Center
AFCMC
   Air Force Contract Maintenance Center
AFCMD
   Air Force Contract Management Division
AFCOMS
   Air Force Commissary Service
AFCOS
   Air Force Combat Operations Staff
AFCPMC
   Air Force Civilian Personnel Management Center
AFCS
   (1) Active Federal Commissioned Service
   (2) Automatic Flight Control System
AFDTC
   Air Force Development Test Center
AFDW
   Air Force District of Washington
AFEMS
   Air Force Equipment Management System
AFESC
   Air Force Engineering and Service Center
AFETAC
   Air Force Environmental Technical Applications Center
AFFDL
   Air Force Flight Dynamics Laboratory
AFFMA
   Air Force Frequency Management Agency
AFFS
   Army Field Feeding System
AFFSA
   Air Force Flight Standards Agency
AFFTC
   Air Force Flight Test Center
AFG
   Auto Force Generator
AFGL
   Air Force Geophysics Laboratory
AFGWC
   Air Force Global Weather Center

AFHRA
   Air Force Historical Research Agency
AFHRL
   Air Force Human Resources Laboratory
AFIA
   Air Force Inspection Agency
AFIC
   (1) Air Force Intelligence Command [established 1991] (USAF)
   (2) American Forces Information Council
AFID
   Anti-Fratricide Identification Device
AFIFIO
   Air Force Information For Industry Office
AFIO
   Association of Former Intelligence Officers
AFIS
   (1) Air Force Intelligence Service
   (2) American Forces Information Services
AFISA
   Air Force Intelligence Support Agency
AFISC
   Air Force Inspection and Safety Center
AFIT
   Air Force Institute of Technology [Wright-Patterson AFB]
AFLC
   Air Force Logistics Command [replaced by AFMC, deactivation 1 July 1992]
AF/LE
   Deputy Chief of Staff (DCS) for Logistics and Engineering (USAF) [See also DCS/LE]
AF/LEE
   Directorate of Engineering and Services (USAF)
AF/LET
   Directorate of Transportation (USAF)
AF/LEX
   Directorate of Plans and Programs (USAF)
AF/LEY
   Directorate of Maintenance and Supply (USAF)
AFLMC
   Air Force Logistics Management Center
AFLSA
   Air Force Legal Services Agency

AFLSC
  Air Force Legal Services Center
AFM
  Air Force Manual
AFMC
  Air Force Materiel Command [established 1992]
AFMEA
  Air Force Management Engineering Agency
AFML
  Air Force Matériels Laboratory
AFMOA
  Air Force Medical Operations Agency
AFMPC
  (1) Air Force Manpower and Personnel Center
  (2) Air Force Military Personnel Center
AF/MPK
  Directorate of Civilian Personnel (USAF)
AF/MPM
  Directorate of Manpower and Organization (USAF)
AF/MPP
  Directorate of Personnel Programs (USAF)
AF/MPX
  Directorate of Personnel Plans (USAF)
AFMSA
  Air Force Medical Support Agency
AFMSS
  Air Force Mission Support System
AFMWRA
  Air Force Morale, Welfare, and Recreation Agency
AFN
  Armed Forces Network [TV-radio]
AFNEWS
  Air Force News Agency
AFNORTH
  Allied Forces Northern Europe (NATO)
AFOE
  Assault Follow-On Echelon [of amphibious force] (USN-USMC)
AFOMS
  Air Force Office of Medical Support
AFOSI
  Air Force Office of Special Investigations

AFOSP
>Air Force Office of Security Police

AFOSR
>Air Force Office of Scientific Research

AFOTEC
>Air Force Operational Test and Evaluation Center

AFP
>Approval for Full Production (USN)

AFPC
>(1) Air Force Personnel Council
>(2) Armed Forces Policy Council

AFPEA
>Air Force Packaging Evaluation Agency

AFPEO
>Air Force Program Executive Office

AFPPS
>American Forces Press and Publication Service

AF/PR
>Deputy Chief of Staff (DCS) for Programs and Evaluation (USAF)

AF/PRI
>Directorate of International Programs (USAF)

AFPRO
>Air Force Plant Representative Office

AF/PRP
>Directorate of Program and Evaluation (USAF)

AFQT
>Armed Forces Qualification Test

AFR
>(1) Advanced Fleet Reactor (USN)
>(2) Air Force Regulation
>(3) Air Force Reserve [also AFRES]

AFRBA
>(1) Air Force Review Boards Agency
>(2) Armed Forces Relief and Benefit Association

AF/RD
>Deputy Chief of Staff (DCS) for Research, Development, and Acquisition (USAF)

AF/RDC
>Directorate of Contracting and Acquisition Policy (USAF)

AF/RDP
>Directorate of Development and Programming (USAF)

AF/RDQ
Directorate of Operational Requirements (USAF)
AF/RDS
Directorate of Command, Control, Communications, and Information (USAF)
AF/RDX
Directorate of Program Integration (USAF)
AFRES
Air Force Reserve [also AFR]
AFROTC
Air Force Reserve Officer Training Corps
AFRPL
Air Force Rocket Propulsion Laboratory
AFRRI
Armed Forces Radiobiology Research Institute
AFRTS
Armed Forces Radio and Television Service
AFS
(1) Active Federal Service
(2) Afloat Prepositioning Ship
AFSA
Air Force Safety Agency
AFSAA
Air Force Studies and Analyses Agency
AFSAC
Air Force Special Activities Center
AFSARC
Air Force Systems Acquisition Review Council
AFSATCOM
Air Force Satellite Communications System
AFSC
(1) Air Force Space Command
(2) Air Force Systems Command [replaced by AFMC]
(3) Armed Forces Staff College
AFSCF
Air Force Satellite Control Facility
AFSCN
Air Force Satellite Control Network
AFSCO
Air Force Security Clearance Office
AFSCP
Air Force Systems Command Pamphlet

AFSINC
Air Force Service Information and News Center
AFSOC
Air Force Special Operations Command [established 1990]
AFSOUTH
Allied Forces Southern Europe (NATO)
AFSPA
Air Force Security Police Agency
AFSPACECOM
Air Force Space Command
AFSTC
Air Force Space Technology Center
AFTAC
Air Force Technical Applications Center
AFTI
Advanced Fighter Technology Integration [F-16] (USAF)
AFV
Armored Fighting Vehicle
AFWAL
Air Force Wright Aeronautical Laboratories
AFWL
Air Force Weapons Laboratory
AF/XO
Deputy Chief of Staff (DCS) for Plans and Operations (USAF)
AF/XOE
Directorate of Electromagnetic Combat (USAF)
AF/XOK
Directorate of Command, Control, and Communication (USAF)
AF/XOO
Directorate of Operations (USAF)
AF/XOS
Directorate of Space (USAF)
AF/XOX
Directorate of Plans (USAF)
AG
Adjutant General (USA)
AGC
Adjutant General's Corps (USA)
AGCW
Autonomously Guided Conventional Weapon
AGE II
Air Ground Engagement System II (USA)

AGL
 Above Ground Level
AGM
 (1) Aircraft Ground Mishap
 (2) Air-to-Ground Missile
AGMC
 Aerospace Guidance and Meteorology Center (USAF)
AGPU
 Aviation Ground Power Unit
AGR
 Active Guard and Reserve (USA)
AGS
 Armored Gun System (USA)
AGSM
 Anti-G Straining Maneuver
AGT
 Advanced Gun Technology
AGW
 Autonomous Guided Weapon (USAF)
AH
 Attack Helicopter (USA)
AHA
 Alert Holding Area (USA)
AHB
 Attack Helicopter Battalion (USA)
AHIP
 Army Helicopter Improvement Program
AHRS
 Altitude and Heading Reference System
AI
 (1) Air Intelligence
 (2) Air Intercept
 (3) Air Interdiction
 (4) Artificial Intelligence
AIA
 Aerospace Industries Association
AIASA
 Annual Integrated Assessment of Security Assistance
AIC
 (1) Airborne Intercept Controller (USN)
 (2) Atlantic Intelligence Center

AID
    Agency for International Development
AIF
    Automated Installation Intelligence File
AIFV
    Armored Infantry Fighting Vehicle
AIK
    Assistance-In-Kind
AIM
    (1) Aircraft Intermediate Maintenance Detachment (USN-USMC)
    (2) Air-Intercept Missile
AIMD
    Aircraft Intermediate Maintenance Department (USN)
AIMI
    Aviation Intensive Management Item (USA)
AIMVAL
    Air Intercept Missile Evaluation (USAF-USN)
AIO
    Air Intelligence Officer [now IO] (USN)
AIP
    Air Independent Propulsion [submarines]
AIRBOC
    Airborne Rapid-Blooming Off-Board Chaff
AIRCENT
    Air Forces, U.S. Central Command [see USAIRCENT]
AIRCOM
    Air Command [Canadian]
AIRSOUTH
    Allied Air Forces Southern Europe (NATO)
AIS
    Automated Information System
AISMO
    Afloat Intelligence System Manager Overview
AIT
    (1) Advanced Individual Training (USA)
    (2) Airborne Integrated Terminal (USAF)
AIWS
    Advanced Interdiction Weapons System (USAF-USN)
AJ
    Anti-Jam
AJCM
    Anti-Jam Control Modem

AL
  Acquisition Logistician
ALB
  Air-Land Battle (USA)
ALC
  Air Logistics Center (USAF)
ALCC
  Airlift Control Center (USAF)
ALCE
  Airlift Control Element (USAF)
ALCM
  Air-Launched Cruise Missile [missile designation AGM-86]
ALD
  Available-to-Load Date [at Port of Embarkation—POE]
ALE
  Automatic Link Establishment
ALF
  Auxiliary Landing Field (USN)
ALFS
  Airborne Low-Frequency Dipping Sonar (USN)
ALICE
  All-purpose, Lightweight, Individual, Carrying Equipment [pack]
ALMAR
  All-Marine (USN-USMC)
ALMC
  Army Logistics Management College
ALMV
  Air-Launched Miniature Vehicle
ALNAV
  All-Navy [message intended for general distribution] (USN)
ALO
  (1) Air Liaison Officer
  (2) Authorized Level of Organization (USA)
ALOC
  (1) Air Lines Of Communication
  (2) Aviation Logistics and Operations Center (USA)
ALP
  Approval for Limited Production (USN)
ALPS
  Accidental Launch Prevention System [ABM concept]
ALS
  Advanced Launch System

ALSE
Aircraft Life Support Equipment (USA)
ALUSNA
U.S. Naval Attaché
ALWT
Advanced Lightweight Torpedo [now Mk 50] (USN)
AMARC
Aerospace Maintenance and Regeneration Center (USAF)
AMB
Air Mission Brief (USA)
AMC
(1) Air Mission Commander (USA)
(2) Air Mobility Command [established 1992] (USAF)
(3) Army Materiel Command (USA)
(4) Army Medical Corps (USA)
AMCCOM
Armament, Munitions, and Chemical Command (USA)
AMCM
Airborne Mine Countermeasures (USN)
AMCS
Aircrew Microclimate Cooling System (USA)
AMCSS
Army Military Clothing Sales Store
AMC-SWA
AMC-Southwest Asia
AMD
Aerospace Medical Division (USAF)
AMDAS
Airborne Mine Detection and Surveillance System (USN)
AMDO
Aeronautical Maintenance Duty Officer (USN)
AMDS
Advanced Mine Detection System (USN)
AME
Average Monthly Earning
AMF
Ace Mobile Force (NATO)
AMG
Aviation Maintenance Group (USA)
AMHS
Automated Message Handling System

AMI
American Military Institute
AMNSYS
Airborne Mine Neutralization System
AMOS
Air Force Maui Optical Station
AMP
(1) Advanced Management Program (USA)
(2) Avionics Modernization Program (USAF)
Amph
Amphibious
AMPHIBIND
Amphibious Warfare Indoctrination
AMPHIBINT
Amphibious Intelligence
AMPHIBPLN
Amphibious Planning
AMPS
Auxiliary Marine Power Source [for submarines]
AMRAAM
Advanced Medium-Range Air-to-Air Missile; missile designation AIM-120A
AMS
Actuation Mine Simulator
AMSAA
Army Material Systems Analysis Agency
AMSC
Army Management Staff College
AMSDL
Acquisition Management System Data List
AMTI
Airborne Moving-Target Indicator
AMW
Amphibious Warfare (USN)
An
Antonov (Russian aircraft designation)
AN/
prefix for U.S. electronic designation systems [e.g., AN/SPS-48; *originally* AN indicated Army-Navy; see table 1]
ANBACIS
Automated Nuclear, Biological, and Chemical Information System

Table 1. U.S. Electronic Designations

Explanation of symbols:

AN/SPG-60

Prefix—Joint service designation	1st symbol—Installation	2nd symbol—Type of equipment	3rd symbol—Purpose	4th symbol—Series
	A = Aircraft B = underwater (submarine) S = Surface ship U = multi-platform W = surface ship and underwater (submarine)	A = invisible light, heat, radiation L = countermeasures P = radar Q = sonar R = Radio S = Special W = Weapon related	D = Direction finder or reconnaissance E = Ejection (e.g. chaff) G = fire control N = Navigation Q = multiple or special purpose R = Receiving, passive detection S = Search W = Weapon control Y = multi-function	[60th series]

SOURCE: Norman Polmar, *The Naval Institute Guide to the Ships and Aircraft of the U.S. Fleet*, 15th ed. (Annapolis: Naval Institute Press, 1993), 515.

ANC
　Army Nurse Corps
ANCC
　Automated Network Control Center
ANCOC
　Advanced NCO Course (USA)
ANDVT
　Advanced Narrowband Digital Voice Terminal
ANG
　Air National Guard (USAF)
ANGB
　Air National Guard Base (USAF)
ANGLICO
　Air-Naval Gunfire Liaison Company (USN-USMC)
ANGOSA
　Air National Guard Operational Support Aircraft (USAF)
ANGSC
　Air National Guard Support Center
ANMCC
　Alternate National Military Command Center
ANVIS
　Aviator's Night Vision Imaging System (USA)
ANZUS
　Australia, New Zealand, and U.S. [alliance]
AO
　(1) Aerial Observer (USA-USMC)
　(2) Air Officer (USMC)
　(3) Area of Operation (USA)
AOA
　(1) Airborne Optical Adjunct (USA)
　(2) Amphibious Objective Area (USN-USMC)
　(3) Angle Of Attack
AOC
　(1) Area of Concentration (USA)
　(2) Army Operations Center (USA)
　(3) Association of Old Crows
　(4) Aviation Officer Candidate (USN)
AOCP
　Aviation Officer Continuation Pay (USN)
AOCS
　Aviation Officer Candidate School (USN)

AOE
　Army Of Excellence
AOR
　Area Of Responsibility
AOSL
　Authorized Organizational Stockage Lists
AOTA
　All-Optical Towed-Array Sonar (USN)
AP
　(1) Acquisition Plan
　(2) Anti-Personnel
　(3) Armor Piercing
AP( )
　Aircraft Procurement (Appropriations); ( ) = A for Army, N for Navy, or AF for Air Force
APAM
　Anti-Personnel/Anti-Matériel (munitions)
APAS
　Alternative Performance Appraisal System
APB
　Acquisition Program Baseline
APC
　Armored Personnel Carrier
APCS
　Approach Power Compensator System (USN)
APDS
　Armor Piercing Discarding Sabot [ammunition]
APF
　Afloat Prepositioning Force
APFSDS
　Armor Piercing Fin Stabilized Discarding Sabot (USA)
APFSDS-T
　Armor Piercing Fin Stabilized Discarding Sabot-Tracer (USA)
APG
　Aberdeen Proving Ground [Maryland] (USA)
APL
　(1) Applied Physics Laboratory [Johns Hopkins University]
　(2) Approved Parts List
APN
　Aircraft Procurement, Navy [funding category]
APO
　Armed Forces Post Office

APOBS
Anti-Personnel Obstacle Breaching System (USA)
APOD
Aerial Port Of Debarkation
APOE
Aerial Port Of Embarkation
APORTS
Aerial Ports and Air Operating Base File
APPN
Appropriation
APRT
Army Physical Readiness Test
APS
(1) Aerial Port Squadrons
(2) Afloat Planning System (USN)
(3) Afloat Prepositioned Ship (USN)
APU
Auxiliary Power Unit
AQAP
Allied Quality Assurance Provision
AQQ
Annual Qualification Questionnaire (USN)
AR
(1) Army Regulation (USA)
(2) Automatic Rifle [obsolete]
ARAP
Army Research Associates Program
ARB
(1) Acquisition Review Board
(2) Air Reserve Base (USAF)
ARBS
Angle Rate Bombing Set
ARC
(1) Air Reserve Components
(2) American Red Cross
ARCENT
Army Forces, U.S. Central Command [see USARCENT]
ARCOM
Army Reserve Command
ARDEC
Army Armament Research, Development, and Engineering Center

AREC
Air Resources Element Coordinator
ARF
(1) Air Refueling Facility
(2) Air Reserve Forces (DOD)
ARFA
Allied Radio Frequency Agency (NATO)
ARFF
Air Reserve Forces Facility
ARG
(1) Air Refueling Group (USAF)
(2) Amphibious Ready Group (USN-USMC)
ARIA
Advanced Range Instrumentation Aircraft (USAF)
ARL
Airborne Recce Low
ARLANT
U.S. Army Atlantic
ARM
Anti-Radiation Missile
Armd
Armored
ARNG
Army National Guard
AROERCEN
Army Reserve Personnel Center
ARPA
Advanced Research Projects Agency [formerly DARPA; changed in 1993]
ARPC
Air Reserve Personnel Center (USAF)
ARPERCENT
Army Reserve Personnel Center
ARRDATE
Arrival Date
ARRS
Aerospace Rescue and Recovery Service (USAF)
ARS
Advanced Rocket System
ARSOC
Army Special Operations Command

ARSOFTF
　Army Special Operations Forces Task Force
ARSTAF
　Army Staff
ARTB
　Advanced Radar Test-Bed
ARTEP
　Army Training and Evaluation Program
Arty
　Artillery
AS-( )*
　Designation for Soviet-Russian Air-to-Surface missile (NATO)
ASA
　Assistant Secretary of the Army
ASAC
　Anti-Submarine Aircraft Control (USN)
ASAF
　Assistant Secretary of the Air Force
ASAF(A)
　Assistant Secretary of the Air Force (Acquisition)
ASAG
　Aegis Surface Action Group
ASAP
　As Soon As Possible [slang]
ASAR
　All Source Analysis System (USA)
ASARC
　Army Systems Acquisition Review Council
ASA (RDA)
　Assistant Secretary of the Army (Research, Development, and Acquisition)
ASARS
　Advanced Synthetic Aperture Radar System (USAF)
ASAS
　All Source Analysis System (USA)
ASAT
　Anti-Satellite
ASB
　Army Science Board
ASBCA
　Armed Services Board of Contract Appeals

ASBPO
Armed Services Blood Program Office
ASBREM
Armed Services Biomedical Research and Evaluation Management Committee
ASC
(1) Advanced Systems Concept (USN)
(2) Army Staff College
ASCM
Anti-Ship Cruise Missile
ASCOMED
Air Service Coordination Office, Mediterranean (USN)
ASD
(1) Aeronautical Systems Division (USAF)
(2) Assistant Secretary of Defense
ASDC
Army Strategic Defense Command
ASD(C^3I)
ASD (Command, Control, Communications, and Intelligence)
ASD(FM&P)
ASD (Force Management and Personnel)
ASD(HA)
ASD (Health Affairs)
ASDIC
British term for sonar[1]
ASD(LA)
ASD (Legislative Affairs)
ASD(PA)
ASD (Public Affairs)
ASD(PA&E)
ASD (Program Analysis and Evaluation)

---

[1]There is considerable confusion over the origins of the term ASDIC. After Winston Churchill used the term in the House of Commons in December 1939, the Admiralty stated that the word was an acronym for *A*llied *S*ubmarine *D*etection *I*nvestigation *C*ommittee, "a body which was formed during the war of 1914–1918, and which organized much research and experiment for the detection of submarines." No committee bearing this name, however, has been found in the Admiralty archives. The term has also been cited as indicating *A*nti-*S*ubmarine *D*ivision, the Admiralty department that sponsored anti-submarine research in World War II.

ASD(P&L)
ASD (Production and Logistics)
ASD(RA)
ASD (Reserve Affairs)
ASDS
Advanced Seal Delivery System [formerly Advanced Swimmer Delivery System] (USN)
ASE
Aircraft Survivability Equipment (USA)
ASEAN
Association of Southeast Asian Nations
ASF
(1) Air-Superiority Fighter (USAF)
(2) Army Stock Fund
ASL
(1) Allowable Supply List
(2) Authorized Stockage List (USA)
ASM
(1) Air-to-Surface Missile
(2) Anti-Ship Missile
(3) Armed Scout Mission
(4) Armored Systems Modernization Program (USA)
(5) Automated Scheduling Message
ASMD
Anti-Ship Missile Defense
ASMRO
Armed Services Medical Regulating Office
ASMS
Advanced Strategic Missile Systems (USAF)
ASN
Assistant Secretary of the Navy
ASN(M&RA)
ASN (Manpower and Reserve Affairs)
ASN(RD&A)
ASN (Research, Development, and Acquisition)
ASO
(1) Aviation Safety Officer (USA)
(2) Aviation Supply Officer (USN)
ASOC
Air Support Operations Center (USA-USAF)
ASP
Ammunition Supply Points (USA)

ASPA
Aircraft Service Period Adjustments (USAF)
A SPEC
System Specification
ASPIS
Advanced Self-Protection Integrated Suite
ASPJ
Airborne Self-Protection Jammer [AN/ALQ-165(V)]
ASR
(1) Acquisition Strategy Report
(1) Air-Sea-Rescue [SAR is now in general use]
(2) Armed Surface Reconnaissance (USN)
ASRAAM
Advanced Short-Range Air-to-Air Missile (missile designation AIM-132)
ASROC
Anti-Submarine Rocket (missile designation RUR-5A) (USN)
ASSETS
Transportation Assets File
AST
(1) Airborne Surveillance Testbed (USA)
(2) Assignment Specific Training
(3) Aviation Selection Test (USN)
ASTOR
Anti-Submarine Torpedo [Mk 45; discarded]
ASTOVL
Advanced Short Take-Off/Vertical Landing Aircraft (USMC)
ASTP
Advanced Space Technology Program
ASUW
Anti-Surface Warfare
ASUWC
Anti-Surface Warfare Commander (USN)
ASW
Anti-Submarine Warfare
ASWC
Anti-Submarine Warfare Commander (USN)
ASWCCCS
ASW Center Command and Control System
ASWIXS
ASW Information Exchange System

ASWM
  ASW Module
ASWOC
  ASW Operations Center
AT
  (1) Annual Training (USA)
  (2) Anti-Tank
ATA
  Advanced Tactical Aircraft [later A-12; *not* Advanced Technology Aircraft] (USN)
ATAC
  Advanced Tank Cannon (USA)
ATACC
  Advanced Tactical Air Command Center (USMC)
ATACMS
  (1) Army Tactical Cruise Missile System [outdated; see (2)]
  (2) Army Tactical Missile System
ATAF
  Allied Tactical Air Force (NATO)
ATAH
  Automatic Target Handoff System (USA)
ATARS
  Advanced Tactical Aerial Reconnaissance System [cancelled 1993] (USAF)
ATAS
  Air-To-Air Stinger (USA)
ATATS
  Air Defense Anti-Tank System [*sic*]
ATB
  Advanced Technology Bomber [now B-2]
ATBM
  Anti-Tactical Ballistic Missile [generally replaced by TBMD in U.S. usage]
ATBMP
  Army Technology Base Master Plan
ATC
  (1) Advanced Tanker/Cargo Aircraft
  (2) Air Traffic Control
  (3) Air Training Command (USAF)
  (4) Air Transportable Clinic (USAF)
ATCALS
  Air Traffic Control/Landing System (USAF)

ATCC
   Anti-Terrorism Coordinating Committee
ATCCS
   Army Tactical Command and Control System
ATD
   Advanced Technology Demonstration
ATD/C
   Aided Target Detection and Classification (USA)
ATE
   Automatic Test Equipment
ATF
   (1) Advanced Tactical Fighter [YF-22/YF-23; *not* Advanced Technology Fighter] (USAF)
   (2) Amphibious Task Force (USN-USMC)
ATGM
   Anti-Tank Guided Missile
ATGW
   Anti-Tank Guided Weapon
ATH
   Air Transportable Hospital (USAF)
ATHS
   (1) Airborne Target Handover System (USA)
   (2) Automatic Target Handoff System (USAF)
ATI
   Airborne Track Illuminator
ATLAS
   Advanced Technology Ladar System (USAF)
ATM
   Anti-Tactical Missile
ATO
   (1) Aircraft Transportation Officer (USN)
   (2) Air Tasking Order
ATP
   (1) Acceptance Test Procedures
   (2) Advanced Tactical Prototype
   (3) Allied Tactical Publication
ATS
   (1) Advanced Tactical Support [aircraft] (USN)
   (2) Aircrew Training System (USAF)
ATSA
   Advanced Tactical Support Aircraft (USN)

ATSD
　Assistant to the Secretary of Defense
ATSD(IO)
　ATSD (Intelligence Oversight)
ATSD(IP)
　ATSD (Intelligence Policy)
ATSE
　Automatic Test Set Support (USA)
ATSO
　Air Traffic Services Organization
ATSS
　Advanced Tactical Support System
ATTD
　Advanced Technology Transition Demonstrator
AU
　Air University (USAF)
AUIB
　Aircrew Uniform-Integrated Battlefield (USA)
AUPC
　Average Unit Procurement Cost
AUS
　Army of the United States [total active Army force]
AUSA
　Association of the U.S. Army
AUTEC
　Atlantic Undersea Test and Evaluation Center (USN)
AUTODIN
　Automatic Digital Network
AUTOSEVOCOM
　Automatic Secure Voice Communication
AUTOVON
　Automatic Voice Network [formerly the principal long-haul, unsecure voice communications network within the Defense Communications System; replaced by Defense Switched Network]
AUV
　Autonomous Underwater Vehicle [unmanned] (USN)
AUW
　All Up Weight
AUX
　Auxiliary

AV
  Audio-Visual
(AV)
  Aviation [USN qualification]
AVCATT
  Aviation Combined Arms Tactical Trainer (USA)
AVF
  All-Volunteer Force
AVGAS
  Aviation Gasoline
AVIM
  Aviation Intermediate-Level Maintenance (USA)
AVLB
  Armored Vehicle Launched Bridge
AVPLAN
  Aviation Plan (USMC)
AVROC
  Aviation Reserve Officer Candidate (USN)
AVSCOM
  (1) Army Aviation Systems Command (USA)
  (2) Aviation Support Command (USA)
AVSTOL
  Advanced VSTOL
AVT
  Automatic Video Tracker
AVUM
  Aviation Unit-Level Maintenance (USA)
AWAC
  Airborne Warning And Control (USAF)
AWACS
  Airborne Warning And Control System (USAF)
AWC
  (1) Air War College (USAF)
  (2) Air Warfare Center (USAF)
  (3) Army War College
AWDS
  Automated Weather Distribution System
AWL
  Administrative Weight Limitation
AWOL
  Absent Without Leave

AWPGM
 Adverse Weather Precision Guided Munition (USAF)
AWS
 (1) Air Weather Service (USAF)
 (2) Amphibious Warfare School (USMC)
 (3) Amphibious Warfighting Seminar
AX
 Advanced attack aircraft (USN)

B

BA
 (1) Budget Activity [funding]
 (2) Budget Authority [funding]
BAA
 Broad Agency Announcement
BAC
 Budgeted Cost at Completion
BAe
 British Aerospace
BAFO
 Best And Final Offer
BAI
 Battlefield Air Interdiction (USA-USAF-USN)
BALTAP
 Baltic Approaches
BAOR
 British Army On the Rhine
BAQ
 Basic Allowance for Quarters
BARCAP
 Barrier Combat Air Patrol (USN)
BAS
 Basic Allowance For Subsistence
BASE-LITE
 Base Imagery Transmission Equipment

BASS
  Ballistic Armor Subsystem (USA)
Bat
  Battleship
BAT
  Brilliant Anti-Tank [missile/submunition]
BATES
  Battlefield-Artillery Engagement System (USA)
BBBG
  Battleship Battle Group [usually BBG] (USN)
BBG
  Battleship Battle Group [sometimes the more awkward BBBG] (USN)
BBLS
  Barrels
BBS
  Brigade/Battalion Simulation (USA)
BB&T
  Blocking, Bracing, and Tie-down (USA)
BCCI
  Base Case Coordinating Instructions
BCD
  Biological/Chemical Detector
BCE
  Baseline Cost Estimate
BCM
  Base Correlation Matrix (USAF)
BCS
  Battery Computer System
BCT
  Bomber Control Team (USAF)
BCTP
  Battle Command Training Program (USA)
BCWP
  Budgeted Cost for Work Performed
BCWS
  Budgeted Cost for Work Scheduled
BDA
  (1) Battle Damage Assessment
  (2) Bomb Damage Assessment
BDAR
  Battle Damage Assessment and Repair (USA)

BDAT
   Battle Damage Assessment Team (USN)
Bde
   Brigade (USA) [see also Brig]
BDFA
   Basic Daily Food Allowance
BDM
   Bunker Defeat Munition (USA)
BDO
   Battle Dress Overgarment (USA)
BDR
   Battle Damage Repair (USA)
BDRP
   Biological Defense Research Program
BDS
   Business Development Specialist
BDU
   Battle Dress Uniform (USA)
Be
   Beriev [Russian aircraft designation]
BEA
   Budget Enforcement Act
BEEF
   Base Emergency Engineering Force (USAF)
BES
   Budget Estimate Submission
BESS
   Basic Enlisted Submarine School (USN)
BFA
   Battlefield Functional Area (USA)
BFIT
   Battle Force Inport Training (USN)
BFM
   (1) Basic Flight Maneuvering (USN)
   (2) Business and Financial Manager
BFT
   Basic Fighter Transition (USAF)
BFTS
   Bomber/Fighter Training System (USAF)
BFV
   Bradley Fighting Vehicle (USA)

BFVS
Bradley Fighting Vehicle System (USA)
BG
Battle Group (USN)
BGAAWC
Battle Group Anti-Air Warfare Coordinator [no longer used] (USN)
BGCTT
Battle Group Commanders Team Training
BGE
Battle Group Exercise (USN)
BGORS
Battle Group Operational Readiness System
BGPHES
Battle Group Passive Horizon Extension System (USN)
BGTT
Battle Group Tactical Training (USN)
BIC
Battlefield Intelligence Coordinator (USA)
BICES
Battlefield Information Collection and Exploitation System (USAF)
BIS
Board of Inspection and Survey (USN)
BIT
Built-In Test
BITE
Built-In Test Equipment
BITS
Base Information Transfer System
BIW
Bath Iron Works [Maine]
Blk
Block
BLOS
Beyond Line-Of-Sight
BLPS
Ballistic Laser Protective Spectacles (USA)
BLS
Beach Landing Site
BLSA
Basic Load Storage Area

BLSSS
 Base-Level Self-Sufficiency Spares (USAF)
BLT
 Battalion Landing Team (USMC)
BLUFOR
 Blue Force
BM/C^3
 Battlefield Management/Command, Control, and Communications (USA)
BMD
 Ballistic Missile Defense
BMDO
 Ballistic Missile Defense Organization [established 1993] (DOD)
BMEWS
 Ballistic Missile Early Warning System
BMO
 Ballistic Missile Office (USAF)
BMS
 Battlefield Maintenance System (USA)
Bn
 Battalion
B/N
 Bombardier/Navigator (USN)
BNCOC
 Basic NCO Course (USA)
BOA
 Basic Ordering Agreement
BOIP
 Basis Of Issue Plan (USA)
BOM
 Bit Oriented Message
bomb/nav
 (1) bombardier/navigator
 (2) bombing/navigation
BOPS
 Billions of Operations Per Second
BOQ
 Bachelor Officer Quarters
BOV
 Board Of Visitors
bp
 between perpendiculars [length; same as waterline]

BP
>Battle Position (USA)

B&P
>Bid and Proposal

BPDMS
>Basic Point Defense Missile System [Sea Sparrow]

BRAC
>Base Closure and Realignment Commission (DOD)

BRC
>Base Recovery Course (USN)

BRDEC
>Belvoir Research, Development, and Engineering Center (USA)

Brig
>Brigade [see also Bde]

BRT
>Bomber Recovery Team (USAF)

B SPEC
>Development and Specification

BSTF
>Base Shop Test Facility (USA)

BSTS
>(1) Base Shop Test Station (USA)
>(2) Boost Surveillance and Tracking System

BT
>Builder's Trials [ships]

BTI
>Balanced Technology Initiative

BTMS
>Battalion Training Management System (USA)

Btry
>Battery

BTS
>Battalion Targeting System

Bty
>Battery [artillery]

BUD/S
>Basic Underwater Demolition/Seal [course] (USN)

BUF
>Big Ugly F——r [*not* Fellow] [slang for B-52 bomber or other large aircraft; more properly BUFF (see below)]

BUFF
>Big Ugly Fat Fellow [slang for B-52 bomber or other large aircraft]

BUMED
  Bureau of Medicine and Surgery (USN)
BuNo
  Bureau Number (USN)
BUPERS
  Bureau of Naval Personnel [formally now the Naval Military Personnel Command]
BUU
  Basic User Unit (USMC)
BVR
  Beyond Visual Range
BW
  Biological Warfare
BWC
  (1) Biological Weapons Command
  (2) Biological Weapons Convention [1972]
BY
  (1) Base Year
  (2) Budget Year
BZ
  Bravo Zulu [signal for "job well done"] (USN)

C

C-1
  Readiness status: fully combat capable
C^2
  Command and Control
C^2I
  Command, Control, and Intelligence Network
C^2S
  Command and Control System
C^3
  Command, Control, and Communications
C^3BM
  C^3 for Battle Management

C^3CM
Command, Control, and Communications Counter-Measures
C^3I
Command, Control, Communications, and Intelligence
C^3IC
Coalition, Coordination, Communications, and Integration Center
C^3MP
Command, Control, and Communications Master Plan
C^3S
Command, Control, and Communications Systems
C^3SYS DIR
Command, Control, Communications Systems Directorate
C-4
Version of Trident missile; missile designation UGM-96A
C^4
Command, Control, Communications, and Computers
C^4I
Command, Control, Communications, Computers, and Intelligence
C^4I^2
Command, Control, Communications, Computers, Intelligence, and Interoperability (USMC)
C^4ICM
C^4I Counter-Measures
C^4S
Command, Control, Communications, and Computer Systems
(C)
Confidential
CA
Civil Affairs
CAB
Combat Aviation Brigade (USA)
CACO
Casualty Assistance Control Officer (USN)
CAD
Computer-Aided Design
CADC
Central Air-Data Computer (USN)
CADNET
Chemical Agent Detector Network (USA)

CADRE
Center for Aerospace Doctrine, Research, and Education
CADS
(1) Combined Air-Defense System
(2) Computer Aided Design System
CADSVAN
Containerized Ammunition Distribution System Van
CAE
(1) Component Acquisition Executive
(2) Computer-Aided Engineering
CAF
(1) Canadian Air Force
(2) Canadian Armed Forces
CAFMS
Computer-Assisted Force Management System
CAFT
Center for Anti-Fratricide Technology (USA)
CAG
(1) Canadian Air Group
(2) Civil Affairs Group
(3) Commander Air Group [obsolete U.S. Navy term but still used as slang for the Commander Carrier Air Wing]
CAI
Combined Arms Initiative (USA)
CAIG
Cost Analysis Improvement Group (DOD)
cal
caliber
CAL
Computer-Aided Logistics
CALL
Center for Army Lessons Learned
CALOW
Contingency Action/Limited Objective Warfare
CALS
(1) Committee on Ammunition Logistics Support
(2) Computer-Aided Acquisition and Logistics Support (DOD)
(3) Computer-Aided Logistics Support (USN)
(4) Computer-Aided Logistics System (USAF)
CAM
(1) Chemical Agent Monitor (USA)
(2) Commercial Assets Mobilization (USN)

(3) Computer-Aided Manufacturing
(4) Crisis-Action Modules
CANA
Convalescent Antidote for Nerve Agent
CAO
Contract Administration Office
CAP
(1) Civil Air Patrol
(2) Combat Air Patrol
(3) Contractor Acquired Property
(4) Crisis Action Plan
CAPS
Conventional Armaments Planning System
CAPTOR
Encapsulated Torpedo [Mk 60 mine]
Car
Carrier [aircraft carrier]
CAR
(1) Chief, Army Reserve
(2) Combined Arms Regiment (USMC)
(3) Command Assessment Review (USAF)
(4) Configuration Audit Review
CARE
Coronary Artery Risk Evaluation (USAF)
CARQUAL
Carrier Qualification (USN)
CART
Combat Assessment of Readiness and Training
CAS^3
Combined Arms and Service Staff School
CAS
(1) Calibrated Air Speed (USN)
(2) Close Air Support
(3) Combined Air Support (USA)
(4) Contract Administration Services
(5) Cost Accounting Standard
CAS/BAI
Close Air Support/Battlefield Air Interdiction
CASCOM
Combined Arms Support Command
CASN
Command Automated Support Network (USMC)

CASREP
   Casualty Report (USN)
CASS
   Command Active Sonobuoy System
CASTEX
   Coordinator ASW Services and Training Exercises
CASTOR
   Corps Airborne Stand-Off Radar [British-US program]
Cat
   Catapult [slang]
CAT
   (1) Computer-Aided Testing
   (2) Crisis Action Team (USAF)
CATCC
   Carrier Air Traffic Control Center (USN)
CATF
   Commander, Amphibious Task Force (USN)
CATM
   (1) Captive Airborne Training Missile
   (2) Computer-Aided Technical Management
CATS
   Combined Arms Training Strategy (USA)
CATT
   Combined Army Tactical Training System
Cav
   Cavalry (USA)
CAVU
   Ceiling and Visibility Unlimited (USN)
CAW
   Carrier Air Wing [proper abbreviation is CVW; see chapter 3]
CAX
   Combined Arms Exercise (USMC)
CB
   Chemical-Biological
CBD
   (1) Chemical-Biological Defense (USA)
   (2) *Commerce Business Daily*
CBDA
   Chemical and Biological Defense Agency (USA)
CBHA
   Cross Border Humanitarian Assistance

CBIAC
Chemical-Biological [Defense] Information Analysis Center
CBM
Confidence-Building Measures
CBO
Congressional Budget Office
CBPO
Consolidated Base Personnel Office (USAF)
CBPS
Chemical Biological Protective Shelter
CBR
(1) Chemical, Biological, Radiological [warfare]
(2) Concurrent Budget Resolution
CBRN
Caribbean Basin Radar Network
CBRS
Concept Based Requirements System (USA)
CBS
Corps Battle Simulation (USA)
CBTDEV
Combat Developer
CBU
Cluster Bomb Unit
CBW
Chemical Biological Warfare
CCA
Carrier Controlled Approach (USN)
CCAF
Community College of the Air Force
CCAPS
Closed Cycle ADCAP Propulsion System [Mk 48 torpedo improvement]
CCB
Configuration Control Board
CCC
CINC Command Center
CCD
Charge Couple Device
CCDR
Contractor Cost Data Reporting
CCF
Central Personnel Security Clearance Facility

CCH
Chief of Chaplains (USA)
CCIP
Computer Calculated Impact Point
CCN
(1) Configuration Change Notice
(2) Contract Change Notice
CCPO
Consolidated Civilian Personnel Office
CCS
(1) Combat Service Support (USMC)
(2) Command and Control Segment (USAF)
CCSS
Command and Control Switching System
CCT
Combat Control Team (USAF)
CCTS
Combat Crew Training Squadron (USAF)
CCTT
Close Combat Tactical Trainer (USA)
CD
Conference on Disarmament
C&D
(1) Command and Decision
(2) Cover and Deception
C-Day
Day on which a deployment operation is to begin
CDBS
Central Data Base Server
CDC
(1) Central Distribution Center (USA)
(2) Combat Development Command (USMC)
(3) Combat Direction Center [formerly CIC] (USN)
CDCO
Combat Direction Center Officer (USN)
CDE
(1) Chemical Defense Equipment
(2) Conference on Disarmament in Europe
CDI
Conventional Defense Improvements
CDIP
Combined Defense Improvement Projects

CDO
> Command Duty Officer

CDR
> (1) Contractor Design Review
> (2) Critical Design Review

CDRL
> Contract Data Requirement List

CDS
> (1) Combat Direction System
> (2) Congressional Descriptive Summary
> (3) Construction Differential Subsidy

CDTF
> Chemical Defense Training Facility

CE
> (1) Combat Element (USMC)
> (2) Command Element (USMC)
> (3) Concept Exploration
> (4) Corps of Engineers (USA)
> (5) Current Estimate

C-E
> Communications-Electronics

CEAC
> Cost and Economic Analysis Center

CEC
> (1) Civil Engineer Corps (USN)
> (2) Cooperative Engagement Capability (USN)

CECOM
> Communications and Electronics Command (USA)

CE/D
> Concept Exploration/Definition Phase

CEE
> (1) Captured Enemy Equipment
> (2) Commercial Equivalent Equipment (USA)

CEF
> (1) Career Executive Force
> (2) Civil Engineering File

CELV
> Complementary Expendable Launch Vehicle

CEM
> Combined Effects Munitions

CEMO
> Command Equipment Management Office (USAF)

CENTAG
Army Group, Central Europe (NATO)
CENTCOM
U.S. Central Command
CENTO
Central Treaty Organization
CEO
Chief Executive Officer
CEOI
Communications-Electronics Operations Instructions (USA)
CEP
(1) Circular Error Probable[2]
(2) Contract Estimating and Pricing
(3) Cooperative Engagement Processor (USN)
CER
Cost Estimating Relationship
CESP
Civil Engineering Support Plan
CESPG
Civil Engineering Support Plan Generator
CETS
Contractor Engineering and Technical Services
CEV
Combat Engineer Vehicle (USA)
CEWI
Combat Electronic Warfare and Intelligence (USA)
CF
Canadian Forces
CFA
Covering Force Area (USA)
CFC
Combined Forces Command [Republic of Korea–US]
CFE
(1) Contractor Furnished Equipment
(2) Conventional Forces in Europe [treaty; 1990]
CFEN
Contractor Furnished Equipment Notice

---

[2]The radius of a circle within which one-half of the missiles/projectiles are expected to fall.

CFIUS
　　Committee on Foreign Investment in the United States
CFM
　　(1) Contractor Financial Management
　　(2) Contractor Furnished Material
CFP
　　Contractor Furnished Property
CFR
　　Code of Federal Regulation
CFRP
　　Carbon Reinforced Plastic
CFS
　　Community, Family, and Soldier Support Command, Korea
CFSF
　　Contract Funds Status Report
CFSR
　　Contractor Funds Status Report
CFT
　　Conformal Fuel Tank
CFV
　　Cavalry Fighting Vehicle (USA)
CG
　　(1) Chairman's Guidance
　　(2) Coast Guard [preferably USCG]
　　(3) Commanding General
CGAS
　　Coast Guard Air Station
CGC
　　Coast Guard Cutter [normally USCGC]
CGD
　　Coast Guard District
CGNSE
　　Changes in Global National Security Environment
CGR
　　Coast Guard Reserve
CGS
　　CONUS Ground Station
CGSC
　　Command and General Staff College (USA)
CH
　　Cargo Helicopter (USA)

CHAALS
 Communication High Accuracy Airborne Location System (USA)
CHAMPUS
 Civilian Health and Medical Program of the Uniformed Services
CHAPS
 Climatic Heat Aircraft Protective Screen (USA)
CHCS
 Composite Health Care System
Chem
 Chemical
CHINFO
 Chief of Information (USN)
CHNAVRES
 Chief of Naval Reserve
CHOP
 Change of Operational Command (USN)
CHSTR
 Characteristic of Transportation Resource [file]
CI
 (1) Civilian Internee
 (2) Configuration Item
 (3) Counterinsurgency
 (4) Counterintelligence
CIA
 Central Intelligence Agency
CIB
 (1) Combat Infantryman Badge
 (2) Command Information Bureau
CIC
 (1) Combat Information Center [now CDC] (USN)
 (2) Combined Intelligence Center
CICA
 Competition In Contracting Act
CICO
 Combat Information Center Officer (USN)
CID
 (1) Commercial Item Description
 (2) Criminal Investigation Division (USA)
CIDC
 Criminal Investigation Command (USA)
CIE
 Clothing and Individual Equipment (USA)

CILMC
Contingency Intermediate-Level Maintenance Center
CILOP
Conversion-In-Lieu Of Procurement (USN)
CIM
(1) Computer-Integrated Manufacture
(2) Corporate Information Management (USN)
CIN
Cargo Increment Number
CINC
(1) Commander in Chief
(2) Commander in Chief [of U.S. unified or specified command; e.g., CINCLANT]
CINCAFLANT
Commander in Chief, Air Force, Atlantic (USAF)
CINCANT
Commander in Chief, Atlantic
CINCEASTLANT
Commander in Chief, Eastern Atlantic (NATO)
CINCENT
Commander in Chief, U.S. Central Command
CINCEUR
Commander in Chief, U.S. European Command
CINCFOR
Commander in Chief, U.S. Forces Command
CINCHAN
Commander in Chief Channel (NATO)[3]
CINCLANT
Commander in Chief, U.S. Atlantic Command [changed to CINCUSACOM in 1993]
CINCLANTFLT
Commander in Chief, U.S. Atlantic Fleet
CINCMAC
Commander in Chief, Military Airlift Command [obsolete]
CINCNORAD
Commander in Chief, North American Air Defense Command [NORAD]

---

[3]Sometimes spelled CINCCHAN; *most* but not all NATO documents use only two Cs. The position was abolished in 1991.

CINCPAC
  Commander in Chief, U.S. Pacific Command
CINCPACFLT
  Commander in Chief, U.S. Pacific Fleet
CINCSAC
  Commander in Chief, U.S. Strategic Air Command [obsolete]
CINCSO
  Commander in Chief, U.S. Southern Command
CINCSOC
  Commander in Chief, U.S. Special Operations Command
CINCSPACE
  Commander in Chief, U.S. Space Command
CINCSTRAT
  Commander in Chief, U.S. Strategic Command
CINCTRANS
  Commander in Chief, U.S. Transportation Command
CINCUSACOM
  Commander in Chief, U.S.A. Command [established in 1993]
CINCUSARRED
  Commander in Chief, U.S. Army Forces, Readiness Command (USA)
CINCUSEUCOM
  Commander in Chief, U.S. European Command
CINCUSNAVEUR
  Commander in Chief, U.S. Naval Forces Europe
CINTEX
  Combined Inport Training Exercise
CIO
  Central Imagery Office [formerly NRO; changed in 1992] (DOD)
CIP
  (1) Component Improvement Program
  (2) Corps of Intelligence Police (USA)
  (3) Critical Intelligence Parameter
CIPA
  Classified Information Procedures Act
CIR
  Continuing Intelligence Requirement
CIS
  (1) Combat Identification System
  (2) Commonwealth of Independent States [formerly USSR]
CISS
  Center for Information Systems Security (DOD)

CITA
   Commercial or Industrial-Type Activities
CITF
   Combat Identification Task Force (USA)
CITV
   Commander's Independent Thermal Viewer
CIWS
   Close-in Weapon System (USN-USCG)
CJCS
   Chairman of the Joint Chiefs of Staff
CJTF
   Commander Joint Task Force
CLF
   (1) Combat Logistics Force (USN)
   (2) Commander, Landing Force (USN-USMC)
CLGP
   Cannon Launched Guided Projectile [now Copperhead]
CLNP
   Connectionless Network Protocol
CLSF
   Combined Logistics Stores Facility
CLSS
   Combat Logistics Support System (USAF)
CLSU
   Communications Security Logistics Support Unit
CLZ
   (1) Craft Landing Zone
   (2) Cushion Landing Zone (USMC)
CM
   (1) Configuration Management
   (2) Contract Management
CMAGTF
   Contingency MAGTF (USMC)
CMC
   (1) Cheyenne Mountain Complex (USAF)
   (2) Commandant Marine Corps
   (3) Command Master Chief (USN)
Cmd
   Command (USA)
CMEF
   Commander, Middle East Force

CMF
　Career Management Field
CMMCA
　Cruise Missile Mission Control Aircraft
CMMP
　Conventional Munitions Master Plan
CMO
　Civil-Military Operations
CMP
　Configuration Management Plan
CMS
　(1) Classified Materials Systems
　(2) Command Management System
CMSA
　Cruise Missile Support Activity
CMTC
　Combined Arms Maneuver Training Center [Hohenfels, Germany] (USA)
CMV
　Combat Mobility Vehicle (USA)
CN
　Counternarcotics
CNA
　Center for Naval Analyses
CNAD
　Conference of NATO Armaments Directors
CNASP
　Chairman's Net Assesment of Strategic Planning
CNET
　Chief of Naval Education and Training
CNGB
　Chief, National Guard Bureau (USA)
CNN
　Cable News Network
CNO
　Chief of Naval Operations
CNP
　Chief of Naval Personnel
CNR
　(1) Chief of Naval Research
　(2) Chief of Naval Reserve
　(3) Combat Net Radio (USA)

CNVEO
Center for Night Vision and Electro-Optics (USA)
Co
Company [see also Coy]
CO
(1) Change Order
(2) Commanding Officer
(3) Contracting Officer
COA
Course Of Action
COB
Colocated Operating Base
COBOL
Common Business Oriented Language [computer language]
COC
(1) Certificate Of Competency
(2) Certification Of Compliance
COCO
Contractor Owned/Contractor Operated [facilities]
COCOM
(1) Combatant Command [command authority]
(2) Coordinating Committee for Multilateral Export Controls
COD
(1) Carrier On-board Delivery (USN)
(2) Cooperative Opportunities Document
CODAG
Combined Diesel And Gas [turbine]
CODOG
Combined Diesel Or Gas [turbine]
CODELAG
Combined Diesel-Electric And Gas [turbine]
COE
(1) Chief Of Engineers (USA)
(2) Common Operating Environemnt
(3) Corps of Engineers (USA)
COEA
Cost and Operational Effectiveness Analysis
COFT
Conduct Of Fire Trainer
COGAG
Combined Gas turbine And Gas [turbine]

COI
Communications Operating Instruction
COIN
Counterinsurgency
COLA
(1) Cost-Of-Living Allowance
(2) Cost-Of-Living Adjustment
COLT
Combat Observation Lasing Team (USA)
COM
(1) Character Oriented Message
(2) Command
(3) Commander
(4) Communications
COMALF
Commander of Airlift Forces (USAF)
COMAO
Combined Air Operations (USAF)
COMD
Command
COMDT
Commandant
COMINT
Communications Intelligence
COMJTF
Commander, Joint Task Force
COMM
(1) Commission
(2) Communications
COMMZ
Communications Zone (USA)
COMO
Combat Oriented Maintenance Organization (USAF)
COMPT
Comptroller
COMPTUEX
Composite Training Unit Exercise (USN)
COMPUSEC
Computer Security
COMSAT
Communications Satellite

COMSEC
　Communications Security
COMTAC
　Commander, Tactical Air Command [obsolete]
COMUSAFSO
　Commander, USAF Southern Command
COMZ
　Communications Zone
CONAG
　Combined Nuclear And Gas [turbine]
CONAS
　Combined Nuclear And Steam [propulsion]
CONOP
　Concept of Operations
CONOPS
　(1) Concept of Operations
　(2) Contingency Operations
CONPLAN
　(1) Concept Plan
　(2) Operation Plan in Concept Format
CONUS
　Continental United States
CONUSA
　Continental U.S. Army
COOP
　(1) Continuity of Operations Planning
　(2) Craft Of Opportunity Program [minesweeper]
COP
　(1) Concept Outline Plan
　(2) Contigency Operation Plan
COPCOM
　Copernicus Common
COR
　Contrating Officer's Representative
CORE
　(1) Contingency Response
　(2) Contingency Response Program of the Navy
CORPSAM
　Corps Surface-to-Air Missile (USA)
CORT
　Escort [as in CORTRON, i.e., Escort Squadron]

COS
(1) Chief Of Staff
(2) Critical Occupational Specialties
COSAL
Coordinated Shipboard Allowance Lists (USN)
COSAN
Combined Steam And Nuclear [propulsion]
COSCOM
Corps Support Command (USA)
COSIRS
Covert Survivable In Weather Reconnaissance and Strike
COSL/COSP
Commander Ocean Systems Atlantic/Pacific
COSO
Combat Oriented Supply Organization (USAF)
COTAC
Copilot/Tactical Coordinator [in S-3 Viking] (USN)
COTAR
Contracting Officer's Technical Representative
COTP
Caption Of The Port (USCG)
COTR
Contracting Officer's Technical Representative
COTS
Commercial Off-The-Shelf
COV
Counter Obstacle Vehicle (USA)
Coy
Company [Co preferred]
CP
(1) Characteristics and Performance
(2) Command Post
CPA
(1) Chairman's Program Assessment [JCS]
(2) Closest Point of Approach
CPAF
Cost-Plus-Award Fee
CPAM
CNO Program Assessment Memorandum (USN)
CPD
Congressional Presentation Document

CP/D
    Cost/Pricing Data
CPE
    Consumer Premise Equipment
CPFF
    Cost-Plus-Fixed Fee
CPFL
    Contingency Planning Facilities List
CPG
    Contingency Planning Guidance
CPI
    Consumer Price Index
CP&I
    Coastal Patrol and Interdiction (USN)
CPIF
    Cost-Plus-Incentive Fee
CPM
    (1) Contractor Performance Measurement
    (2) Critical Path Method
CPMC
    Civilian Personnel Management Center (USAF)
CPO
    (1) Chemical Protective Overgarment (USA)
    (2) Civilian Personnel Office
CPR
    Cost Performance Report
CPS
    (1) Collective Protection System (USN)
    (2) Competitive Prototyping Strategy
CPSR
    (1) Contract Procurement System Review
    (2) Contract Purchasing System Review
CPU
    Central Processing Unit
CPX
    Command Post Exercise
CQ
    (1) Carrier Qualification (USN)
    (2) Change of Quarters
CQB
    Close Quarter Battle

CQT
  Command Quality Team
CR
  (1) Continuing Resolution
  (2) Cost Reimbursement
CRA
  Continuing Resolution Authority
CRAF
  Civil Reserve Airlift Fleet
CRAFTS
  Civil Reserve Auxiliary Fleet Ships
CRAG
  Contractor Risk Assessment Guide
CRC
  (1) Continental U.S. Replacement Center (USA)
  (2) Control and Reporting Center (USA)
  (3) CONUS Replacement Center
CRDA
  Cooperative Research and Development Agreement
CRG
  Cryptologic Readiness Group
CRI
  CHAMPUS Reform Initiative
CRISD
  Computer Resource Integrated Support Document
CRITIC
  Critical Intelligence Report
CRITICOMM
  Critical Intelligence Communication System
CRLCMP
  Computer Resource Life Cycle Management Plan
CRP
  Control and Reporting Post (USAF)
CRRC
  Combat Rubber Raiding Craft
CRS
  (1) Computer-Assisted Requirement Catalog
  (2) Congressional Research Service [Library of Congress]
CRSG
  Contractor Risk Assessment Guides
CRSP
  Combat Ready Storage Program

CRT
Cathode Ray Tube
CRU
Cruiser [as COMCRUGRU]
CRWG
Computer Resource Working Group
CS
Combat Support
CSA
(1) Chief of Staff of the Army
(2) Combat Systems Assessment (USN)
(3) Compliance Schedule Approval
CSAF
Chief of Staff Air Force
CSAR
Combat Search And Rescue
CSAT
Combat Systems Assessment (USN)
CSBM
Confidence and Security-Building Measures (NATO)
CSC
(1) Combined Service Command
(2) Command Senior Chief (USN)
(3) Conventional Systems Committee (DOD)
CSCE
Conference on Security and Cooperation in Europe (NATO)
CSCI
Computer Software Configuration Item
C/SCSC
Cost/Schedule Control System Criteria
CSDP
Chemical Stockpile Disposal Program
CSF
Contingency Support Force (USAF)
CSG
Cryptologic Support Group
CSH
Combat Support Hospital
CSI
Competitive Strategies Initiative

CSIS
Center for Strategic and International Studies [Washington, D.C.; formerly Georgetown CSIS]
CSM
Command Sergeant Major
CSMC
Combat System Maintenance Central (USN)
CSMNS
Combat Swimmer Mine Neutralization System
CSMP
Current Ship's Maintenance Project (USN)
CSOA
Combined Special Operations Area
CSOC
Consolidated Space Operations Center (USAF)
CSOM
Computer Software Operator's Manual
CSOSS
Combat System Operational Sequencing System (USN)
CSP
Common Signal Processor
CSPA
CINC's Strategic Priorities Assessment
CSPAR
CINC
Preparedness Assessment Report
CSR
Controller Supply Rate (USA)
CSRL
Common Stategic Rotary Launcher
CSRR
Combat Systems Readiness Review
CSRS
Common Source Routing File
CSS
(1) Central Security Service
(2) Combat Service Support
(3) Communication Support Service
(4) Contractor Support Services
CSSA-1
(1) CENTAF Supply Support Activity (USAF)
(2) Combat Service Support Area (USMC)

CSSCS
Combat Service Support Control System (USA)
CSSD
Combat Service Support Detachment (USMC)
CSSE
Combat Service Support Element (USMC)
CSSR
Cost/Schedule Status Report
CSW
Conventional Standoff Weapon
CSWS
Corps Support Weapon System
CT
(1) Cased Telescoped [type of ammunition]
(2) Counter-Terrorism
CTA
Common Table of Allowance
CTAPS
Contingency Tactical Air Control Automated Planning System
CTASC
Corps/Theater ADP and Service Center (USA)
CTB
Comprehensive Test Ban [nuclear weapon tests]
CTC
Combat Training Center (USA)
CTEA
Cost and Training Effectiveness Analysis
CTF
(1) Combined Task Force (USN)
(2) Commander Task Force
CTG
Commander Task Group
CTIS
Central Tire Inflation System (USA)
CTJTF
Counter-Terrorist Joint Task Force
CTO
Communication Technician Operator
CTOL
Conventional Take-Off and Landing
CTS
Contact Test Set (USA)

CTT
  Coordination and Training Team [SOF]
CTTG
  Counter Targeting (USN)
CTU
  Commander Task Unit
CUDIXS
  Common User Data Information Exchange System
CV
  Cargo Variant [LSD 41 variant]
CVBG
  Aircraft Carrier Battle Group (USN)
CVC
  Combat Vehicle Crewman
CVG
  Carrier Air Group [no longer used]
CVIC
  Carrier Intelligence Center (USN)
C-V-P
  Cost-Volume-Profit
CW
  (1) Chemical Warfare
  (2) Chemical Weapons
CWBS
  Contract Work Breakdown Structure
CWC
  (1) Chemical Weapons Convention
  (2) Composite Warfare Commander (USN)
CWOSM
  Composite Warfare Oceanographic Support Module (USN)
CWS
  Chemical Warfare Service (USA)
CWTPI
  Conventional Weapons Tactical Proficiency Inspection (USN)
CY
  (1) Calendar Year
  (2) Current Year
CZ
  Convergence Zone [acoustic]

D

D-5
Version of Trident missile [missile designation UGM-133A]
DA
(1) Decision Analysis
(2) Department of the Army
(3) Developing Activity
(4) Developing Agency
(5) Direct Action [special operations forces]
DAACM
Direct Airfield Attack Combined Munition (USAF)
DAB
Defense Acquisition Board
DAC
Defense Acquisition Circular
DACOWITS
Defense Advisory Committee On Women In The Service
DACT
Dissimilar Air Combat Training (USN)
DAE
Defense Acquisition Executive
DAES
Defense Acquisition Executive Summary
DAF
Department of the Air Force
DALSO
Department of the Army Logistics Staff Officer
DAM
Decontaminating Agent, Multipurpose
DA&M
Director of Administration and Management (DOD)
DAMA
Demand Assigned Multiple Access [multiplexing]
DAP
Department of the Army Publication
DAR
Defense Acquisition Regulation
DARC
Defense Acquisition Regulatory Council

DARCOM
Development and Readiness Command (USA)
DARE
Drug Abuse Resistance Education (DOD)
DARMS
Developmental Army Mobilization System
DARPA
Defense Advanced Research Projects Agency
DARS
Daily Aerial Reconnaissance and Surveillance
DART
Downed Aircraft Recovery Team (USA)
DAS
(1) Direct Air Support
(2) Director of the Army Staff
DASC
(1) Department of the Army System Coordinator
(2) Direct Air Support Center
DASC-A
Direct Air Support Center—Airborne
DASD
Deputy Assistant Secretary of Defense
DASH
Drone Anti-Submarine Helicopter [discarded]
DAT
Deployment Action Team (USA)
DAU
Defense Acquisition University
DAWIA
Defense Acquisition Work Force Improvement Act
DBDD
Data Base Design Document
DBDU
Desert Battle Dress Uniform
DBOF
Defense Business Operating Fund
DC
Depth Charge
DCA
(1) Damage Control Assistant (USN)
(2) Defense Communications Agency [obsolete; see DISA]
(3) Defense Cooperation Account

(4) Defensive Counter-Air
(5) Dual-Capable Aircraft
DCAA
Defense Contract Audit Agency
DCAG
Deputy Air Wing Commander [no longer used] (USN)
DCAS
Defense Contract Administration Services
DCI
Director of Central Intelligence
DCID
Director of Central Intelligence Directive
DCIMI
Defense Council on Integrity and Management Improvement
DCL
Direct Communication Link
DCM
(1) Deputy Commander for Maintenance (USAF)
(2) Director of Civilian Marksmanship
DCMAO
Defense Contract Management Area Operation
DCMC
Defense Contract Management Command
DCMR
Defense Contract Management Regions
DCNO
Deputy Chief of Naval Operations
DCO
Deputy Commander for Operations (USAF)
DCP
(1) Decision Coordinating Paper [obsolete]
(2) Director of Air Campaign Plans [CENTCOM]
DCR
Deputy Commander for Resources (USAF)
DCS
(1) Defense Communications System
(2) Defense Courier Service
(3) Deputy Chief of Staff
(4) Digital Computer System
DCSI
Deputy Chief of Staff for Intelligence (USA)

DC/S(I&L)
Deputy Chief of Staff, Installations and Logistics (USMC)
DCSINT
Deputy Chief of Staff for Intelligence (USA)
DCS/LE
Deputy Chief of Staff for Logistics and Engineering [see also AF/LE] (USAF)
DCSLOG
Deputy Chief of Staff for Logistics (USA)
DCSOPS
Deputy Chief of Staff for Operations and Plans (USA)
DCSPER
Deputy Chief of Staff for Personnel (USA)
DCS RD&S
Deputy Chief of Staff for Research, Development, and Studies (USMC)
DCT
Digital Communications Terminal
DCTN
Defense Commercial Telecommunications Network
DCU
Deployment Control Unit (USA)
D-Day
Day on which an operation begins
DDC
Data Distribution Center
DDDR&E (T&E)
Deputy Director, Defense Research and Engineering (Test and Evaluation) (DOD)
DD/EFT
Direct Deposit/Electronic Funds Transfer
DDI
(1) Deputy Director for Intelligence [CIA]
(2) Director of Defense Information
DDN
Defense Data Network
DDO
Directorate of Operations [Central Intelligence Agency]
DDR&E
Director, Defense Research and Engineering (DOD)

DDS
- (1) Data Distribution System (USN)
- (2) Deep Dive System (USN)
- (3) Defense Dissemination System
- (4) Doctor of Dental Surgery
- (5) Dry Deck Shelter (USN)

DDT&E
Director, Defense Test and Evaluation

DDV
Destroyer Variant [Surface Warfare study, USN]

DE
- (1) *Defense Electronics* [magazine]
- (2) Directed Energy [weapons]

DEA
Drug Enforcement Administration

DECA
Defense Commissary Agency

DECCO
Defense Commercial Contracting Office

DECM
- (1) Deception Electronic Countermeasures
- (2) Defensive Electronic Countermeasures

DEERS
Defense Enrollment Eligibility Reporting System

DEFCON
Defense Condition [the higher number indictes a higher state of military readiness; numbered from 1, the highest, through 5]

DEFSMAC
Defense Special Missile and Astronautics Center

DEMNS
Distributed Explosive Mine Neutralization System

DEM/VAL
Demonstration/Validation Phase

DENTAC
Dental Activities (USA)

Dep
Deputy

DEP
- (1) Data Entry Panel
- (2) Defense Enterprise Program

DEPMEDS
Deployable Medical Systems

DEPSECDEF
   Deputy Secretary of Defense
DERA
   Defense Environmental Restoration Account
DERP
   Defense Environmental Restoration Program
DES
   Destroyer [as in COMDESRON]
DESC
   Defense Electronic Supply Center
DESCOM
   Depot System Command (USA)
DESRON
   Destroyer Squadron
DEST
   Destination
Det
   Detachment
DET
   Distributed Explosive Technologies
DEW
   (1) Directed-Energy Weapon
   (2) Distant Early Warning
DF
   Direction Finding [see also HF/DF]
DFARS
   Defense FAR Supplement
DFAS
   Defense Finance and Accounting Service
DFE
   Division Force Equivalent
DFR/E
   Defense Fuel Region/Europe
DFRIF
   Defense Freight Railway Interchange Fleet
DFM
   Diesel Fuel, Marine (USN)
DFRIF
   Defense Freight Railway Interchange Fleet
DFR/ME
   Defense Fuel Region/Middle East

DFSC
Defense Fuel Supply Center
DFSP
Defense Fuel Support Point
DFT
Deployment For Training
DG
Defense Guidance [now Defense Planning Guidance (DPG)]
DGL
Distinguished Guest Lecturer
DGPS
Differential Global Positioning System
DGSC
Defense General Supply Center
DI
Drill Instructor
DIA
Defense Intelligence Agency
DIAC
Defense Intelligence Analysis Center
DIB
Defense Industrial Base
DIC
Dependency and Indemnity Compensation
DICASS
Directional Command Active Sonobuoy System [sonobuoy]
DID
Data Item Description
DIDS
Defense Integrated Data System (USA)
DIET-COLA
Half the rise of the CPI for COLA
DIFAR
Directional Finding And Ranging [sonobuoy]
DIMA
Drilling Individual Mobilization Augmentation (USA)
DINET
Defense Industrial-base Network [now PROBASE]
DINFOS
Defense Information School
DIO
Defense Intelligence Officer

DIP
　Defense Intelligence Plan
DIPEC
　Defense Industrial Plant Equipment Center
DIR
　Director
DIRLAUTH
　Direct Liaison Authorized
DIS
　Defense Investigation Service
DISA
　(1) Defense Information Services Agency
　(2) Defense Information Systems Agency [formerly DCA]
DISAM
　Defense Institute of Security Assistance Management
DISC^4
　Director of Information Systems for Command, Control, Communications, and Computers (USA)
DISCO
　Defense Industrial Security Clearance Office
DISCOM
　Division Support Command (USA)
DISN
　Defense Information System Network
DISNET
　Defense Integrated Secure Network
DISSP
　Defense Information Systems Security Program
Div
　Division (USA)
DIV
　Division [suffix; e.g. CARDIV] (USN)
DivArty
　Division Artillery (USA)
DKIE
　Decontamination Kit Individual Equipment (USA)
DLA
　Defense Logistics Agency
DLAM
　Defense Logistics Agency Manual
DLEA
　Drug Law Enforcement Agencies

D LEVEL
  Depot Level of Maintenance
DLI
  Deck-Launched Interceptor (USN)
DLIR
  Down-Looking Infrared
DLQ
  Deck Landing Qualification (USN)
DLR
  Depot-Level Repairable (USN)
DLSA
  Defense Legal Service Agency
DM
  Director of Management (USA)
DMA
  Defense Mapping Agency
DMAAC
  Defense Mapping Agency Aerospace Center
DMD
  Doctor of Dental Medicine
DME
  Distance Measuring Equipment (USN)
DMFO
  Defense Medical Facilities Office
DMIS
  Defense Medical Information System
DML
  Depot Maintenance Level
DMMIS
  Depot Maintenance Management Information System (USAF)
DMMS
  Depot Maintenance Management System (USMC)
DMO
  Defense Mobilization Order
DMPI
  Designated Mean Point of Impact
DMR
  (1) Defense Management Report
  (2) Defense Management Review
DMRD
  Defense Management Review Decision

DMRIS
Defense Medical Regulating Information System
DMS
(1) Defense Materials System
(2) Defense Messaging System
DMSA
Defense Medical Support Activity
DMSP
Defense Meteorological Support Program [originally Defense Meteorological *Satellite* Program]
DMSQ
Duty Military Occupational Specialty Qualified (USA)
DMSSC
Defense Medicine System Support Center
DMTI
Digital Moving Target Indicator
DMZ
Demilitarized Zone
DNA
Defense Nuclear Agency
DNBI
Disease and Non-Battle Injury
DND
Department of National Defence [Canada]
DNFSB
Defense Nuclear Facilities Safety Board
DNI
Director of Naval Intelligence
DNR
Director of the Naval Reserve
DOD
Department Of Defense [also DoD; all capitals preferred]
DODAAC
DOD Activity Address Code
DODD
(1) DOD Directive
(2) DOD Document
DODDS
DOD Dependents School
DODEX
DOD Intelligence Information System Extension

DODI
DOD Instruction
DODIC
DOD Identification Code
DODIIS
Department of Defense Intelligence Information System
DODISS
DOD Index of Specifications and Standards
DOD-JIC
DOD Joint Intelligence Center
DODMERB
DOD Medical Examination Review Board
DOE
Department of Energy
DON
Department of the Navy
DOPMA
Defense Officer Personnel Management Act
DOS
(1) Days of Supply
(2) Department of State
DOT
Department of Transportation
DOT&E
Director, Operational Test and Evaluation (DOD)
DOW
Died of Wounds
DP
(1) Data Processing
(2) Decision Package
(3) Development Plan
(4) Development Proposal
(5) Dual Purpose [gun suitable for use against air and surface targets]
DPA
Defense Production Act
DPACT
Defense Policy Advisory Committee on Trade
DPC
Defense Panning Committee (NATO)
DPESO
DOD Product Engineering Services Office

DPG
    Defense Planning Guidance [formerly Defense Guidance (DG)]
DPICM
    Dual Purpose Improved Conventional Munition (USA)
DPM
    Deputy Program Manager
DPML
    Deputy Program Manager for Logistics
DPPO
    Development Production Prove Out (USA)
DPRB
    Defense Planning and Resources Board [formerly Defense Resources Board (DRB)]
DPRO
    Defense Plant Representatives Office
DPS
    (1) Decision Package Sets
    (2) Defense Priorities System
    (3) Defense Protective Service
DPSC
    Defense Personnel Support Center
DPSM
    Dual Purpose Submunition (USA)
DQT
    Division Quality Team
DRB
    Defense Resources Board [now Defense Planning and Resources Board (DPRB)]
DRMO
    Defense Reutilization and Materials Organization
DR Plot
    Dead Reckoning Plot (USN)
DRPM
    Direct Reporting Program Manager (USN)
DRU
    Direct Reporting Units (USAF)
DS
    Direct Support
DSA
    (1) Defense Special Assessment [Defense Intelligence Agency]
    (2) Defense Supply Advisor

DSAA
　Defense Security Assistance Agency (DOD)
DSARC
　Defense Systems Acquisitions Review Council
DSB
　Defense Science Board
DSC
　Defense Space Council
DSCS
　Defense Satellite Communications System
DSCSOC
　Defense Satellite Communications Systems Operations Center
DSL
　Chinese-made integrated multisensor firing mechanism
DSMAC
　Digital Scene-Matching Area Correlation
DSMC
　Defense Systems Management College
DSN
　Defense Switched Network
DSNET
　Defense Secure Network
DSP
　Defense Support Program
DSSCS
　Defense Special Security Communications Systems
DSSP
　Defense Standardization and Specification Program
DST
　(1) Decision Support Template
　(2) Destructor [type of mine] (USN)
DSTP
　Director of Strategic Target Planning (USAF)
DT
　Developmental Test
DTC
　(1) Defense Trade Controls [office of]
　(2) Design-To-Cost
DTC-2
　Desktop Computer

DTCN
   Direction Technique des Constructions Navales [French naval design bureau]
DT&E
   Development, Test, and Evaluation
DTED
   Digital Terrain Elevation Data
DTG
   Date-Time Group
DTIC
   Defense Technical Information Center [Defense Logistics Agency]
DTLCC
   Design To Life-Cycle-Cost
DTMB
   David Taylor Model Basin [now David Taylor Research Center] (USN)
DTP
   Defense Trade Policy [office of]
DTSA
   Defense Technology Security Administration
DTUPC
   Design To Unit Production Cost
DUI
   Distinctive Unit Insignia
D/V
   Demonstration/Validation Phase
DVITS
   Digital Video Imagery Transmission System
DVM
   Doctor of Veterinary Medicine
DVQ
   Distinguished Visitor Quarters
DW
   Deep Water
DWT
   Deadweight Tonnage [the ship's cargo-carrying capability]
DY
   Dockyard [not U.S. term]
DZ
   Drop Zone [airborne]

E

"E"
> Excellence [award] (USN)

EA
> (1) Engagement Area (USA)
> (2) Environmental Assessment (EPA)
> (3) Evolutionary Acquisition

EAC
> (1) Eastern Area Command
> (2) Echelons Above Corps (USA)
> (3) Economic Adjustment Committee
> (4) Estimated Cost At Completion

EAC COMM
> Echelon Above Corps Communications (USA)

EAD
> (1) Earliest Arrival Date [at POD]
> (2) Echelons Above Division (USA)

EAM
> Emergency Action Message

EAOS
> End of Active Obligated Service (USN)

EAP
> Emergency Action Plan

EAP-JCS
> Emergency Action Procedure of the Joint Chiefs of Staff

EAPROM
> Electronically Alterable Programmable Read-Only Memory

EB
> Electric Boat [also EB/GD (Division of General Dynamics Corp.; formerly Electric Boat Company]

EBC
> Echelons Below Corps (USA)

EC
> (1) Electronic Combat
> (2) European Community

ECAC
> Electromagnetic Compatibility Analysis Center (USAF)

ECAMP
 Environmental Compliance Assessment and Management Program (USAF)
ECC
 (1) Evacuation Coordination Center
 (2) Executive Communications and Control
ECCM
 Electronic Counter-Countermeasures
ECI
 Extension Course Institute (USAF)
ECM
 Electronic Countermeasures
ECMO
 ECM Officer
ECP
 (1) Engineering Change Proposal
 (2) Enlisted Commissioning Program (USN)
ECR
 Electronic Combat Reconnaissance
ED
 (1) Engineering Development (USA)
 (2) Engineering Duty (USN)
EDA
 Excess Defense Articles
EDC
 (1) Educational Development Center (USAF)
 (2) Estimated Date of Completion of Loading [at POE]
EDD
 (1) Earliest Delivery Date (USN)
 (2) Estimated Departure Date
EDI
 Electronic Data Intelligence
EDM
 Engineering Development Model
EDP
 Emergency Defense Plan
EDP/E
 Electronic Data Processing/Equipment
E&E
 Evasion and Escape

EEAP
  Enlisted Education Advancement Program (USN)
EEC
  European Economic Community
E^3
  Electromagnetic Environmental Effects
EEFI
  Essential Elements of Friendly Information
E^2I
  Endoatmospheric/Exoatmospheric Interceptor (USA)
EEI
  Essential Elements of Information
EEPROMS
  Erasable Electronically Programmable Read-Only Memory
EEZ
  Exclusive Economic Zone
EFA
  European Fighter Aircraft
EFIS
  Electronic Flight Information System
EFS
  Enhanced Flight Screening
EFVS
  Electronic Fighting Vehicle System (USA)
EGT
  Exhaust-Gas Temperature
EHF
  Extremely High Frequency [see table 2]
ehp
  equivalent horsepower
EIA
  Environmental Impact Assessment
EIC
  Equipment Identification Code
EIR
  Equipment Improvement Recommendation (USA)
EIS
  Environmental Impact Statement (EPA)
ELANT
  East Atlantic Satellite
ELF
  Extremely Low Frequency

Table 2. Electromagnetic Frequency Band Designations

Current Frequency Designations Used by USA and NATO	A	B	C	D	E	F	G	H	I	J	K	L	M
Wavelength (cm)	300 200 150	100 75	60 50 40	30 20 15	10		6 5	3.75 3		2 1.5	1 0.75	0.6 0.5	0.4 0.3
Frequency (GHz)	0.1 0.15 0.2	0.3 0.4	0.5 0.6 0.75	1 1.5	3		5 6	8.0 10		15 20	30 40	50 60 70	100
Previous Frequency Designations	VHF	UHF	UHF	L	S	S	C	C	X	K_u	K	K_a	Millimeter
Frequency Designations (WW II)		P	L	L	S	S	C	X	X	K	K	Q V	

SOURCE: Norman Polmar, *The Ships and Aircraft of the U.S. Fleet*, 13th ed. (Annapolis: Naval Institute Press, 1984), 487.

ELINT
Electronic Intelligence
ELS
Emitter Location System
ELT
English Language Training
ELV
Expendable Launch Vehicle
ELW
Electronic Warfare [EW preferred]
EM
Electromagnetic
EMC
(1) Electromagnetic Compatibility
(2) Executive Management Course
EMCON
Emission Control
EMD
Engineering and Manufacturing Development [see also FSD]
EMDP
Engine Model Derivative Program (USAF)
EMI
(1) Electromagnetic Interference
(2) Extra Military Instruction (USN)
EMIS
Electromagnetic Isotope Separation
EML
Environmental and Morale Leave
EMO
(1) Electronics Material Officer
(2) Environmental Management Office
EMP
Electromagnetic Pulse
EMRLD
Excimer Repetitively Pulsed Laser Device (USA)
EMSL
Environmental and Molecular Science Laboratory
EMSP
Enhanced Modular Signal Processor (USN)
EMTI
Enhanced Moving Target Indicator (USAF)

ENCOM
　Engineer Command (USA)
ENCORE
　Enlisted Navy Career Options for Reenlistment
Engr
　Engineer
ENSCE
　Enemy Situation Correlation Element (USAF)
ENWGS
　Enhanced Naval Warfare Gaming System
EO
　(1) Electro-Optical
　(2) Executive Order
EOA
　Early Operational Assessment
EOB
　Electronic Order of Battle
EOCM
　Electro-Optical Countermeasures
EOD
　Explosive Ordnance Disposal
EODTECHCTR
　Explosive Ordnance Disposal Technical Center
EOH
　Equipment-On-Hand (USA)
EOM
　Expendable Ordnance Management (USN)
EOQ
　Economic Ordering Quantity
EOSAT
　Earth Observable Satellite Corporation
EOSS
　Engineering Operational Sequencing System (USN)
EP
　Estimated Position (USN)
EPA
　(1) Economic Price Adjustment
　(2) Environmental Protection Agency
　(3) Extended Planning Annex
EPDS
　Electronic Processing and Dissemination System

EPG
European Participating Governments
EPIC
El Paso Intelligence Center
EPLRS
Enhanced Position Location Reporting System
EPMS
Enlisted Personnel Management System (USA)
EPP
Emergency Power Package (USN)
EPQ
Engineering Qualification Trials
EPROM
Erasable Programmable Read-Only Memory
EPU
Emergency Power Unit (USN)
EPUU
Enhanced PLRS User Unit (USA)
EPW
Enemy Prisoners of War
ERAPS
Expendable Reliable Acoustic Path Sonobuoy
ERCS
Emergency Rocket Communications System [Minuteman II missile]
ERFS
Extended-Range Fuel System
ERINT
Extended-Range Interceptor Technology [missile] (USA)
ERIS
Exoatmospheric Reentry Vehicle Intercept Subsystem (USA)
ERO
Equipment Repair Order (USMC)
EROSAT
ELINT Ocean Reconnaissance Satellite
ERP
(1) Emergency Relocation Point
(2) Expanded Relations Program (USA)
ERT
Execution Reference Time
ESA
European Space Agency

ESC
- (1) Electronic Security Command
- (2) Executive Steering Committee

ESD
- Electronics Systems Division (USAF)

ESF
- Economic Support Fund

ES&H
- Environmental, Safety, and Occupational Health

ESI
- Essential Sustainment Items

ESKE
- Enhanced Station-Keeping Equipment (USAF)

E-SLATS
- Executive Strike Leader Attack Training School

ESM
- (1) Electronic Support Measures
- (2) Electronic Surveillance Methods

ESMC
- Eastern Space and Missile Center (USAF)

E SPEC
- Material Specification

ESS
- External Stores Support (USA)

ESSM
- Evolved Sea Sparrow Missile

EST
- Essential Subjects Test (USMC)

ESWS
- Enlisted Surface Warfare Specialist (USN)

ET
- (1) Electro-Thermal
- (2) Emerging Technology

ETA
- Estimated Time of Arrival

ETAC
- Enlisted Tactical Application

ETC
- Electro-Thermal Chemical [gun]

ETR
- Estimated Time to Repair

ETS
    (1) European Troop Strength
    (2) Expiration of Term of Service
ETUT
    Enhanced Tactical User Terminal
EUCOM
    U.S. European Command
EUCOMM-Z
    U.S. European Command Communications Zone
EUR
    Europe
EUSA
    Eighth United States Army
EUSC
    Effective U.S. Control [merchant shipping]
EVAC
    Evacuation
EVS
    Electro-Optical Viewing System
EW
    Electronic Warfare [see also ELW]
EWC
    Electronic Warfare Coordinator
EWCM
    Electronic Warfare Coordination Module
EWO
    (1) Electronic Warfare Officer
    (2) Emergency War Order
EWS
    (1) Electronic Warfare Supervisor (USN)
    (2) Enlisted Surface Warfare Specialist
EWSE
    Electronic Warfare Support Element (USA)
EXCAP
    Expanded Capability
EXOS
    Executive Office of the Secretary
EXPO
    Extended-range Poseidon [missile, changed to Trident I]

F

FA
Field Artillery
FAA
(1) Federal Aviation Administration
(2) Fleet Air Arm [British]
(3) Foreign Assistance Act [1961]
(4) Forward Assembly Area (USA)
(5) Functional Area Analyses (USMC)
FAAD
Forward Area Air Defense (USA)
FAAD C^2
Forward Area Air Defense Command and Control System (USA)
FAAD C^2I
Forward Area Air Defense Command, Control, and Intelligence (USA)
FAADS
Forward Area Air Defense System (USA)
FAAR
Forward Area Alerting Radar
FAAS
Field Artillery Ammunition Support (USA)
FAASV
Field Artillery Ammunition Support Vehicle
FAAWC
Force Anti-Air Warfare Commander [Royal Navy]
FAC
(1) Fast Attack Craft [not USN ship designation]
(2) Federal Acquisition Circular
(3) Forward Air Control
(4) Forward Air Controller
FAC(A)
Forward Air Control (Airborne) (USMC)
FACP
Forward Air Control Post
FACS
Fleet Area Control and Surveillance Facility (USN)

FAD
  (1) Feasible Arrival Date
  (2) Force Activity Designator
  (3) Forward Area Defense
FADR
  Forward Area Demagnetizing Range
FAE
  Fuel-Air Explosive
FAIR
  Fleet Air (USN)
FAISS-E
  FORSCOM Automated Intelligence Support System, Enhanced (USA)
FAMMO
  Full Ammo (USN)
FAO
  Finance and Accounting Office
FAPES
  Force Augmentation Planning and Execution System
FAR
  (1) Federal Acquisition Regulation
  (2) Federal Acquisition Service
FARE
  Forward Area Refueling Equipment
FARP
  Forward Arming and Refueling Point
FAS
  Federation of American Scientists
FASFAC
  Fast Forward-Air-Control (USMC)
FAST
  (1) Fleet Anti-terrorist Security Team (USMC)
  (2) Forward Area ID and TRAP Broadcast
  (3) Forward Area Support Team (USA)
FASTCAL
  Field Assistance Support Team for Calibration
FASTT
  Fleet All-Source Tactical Terminal
FAT
  (1) Factory Acceptance Test
  (2) Fatalities
  (3) First Article Testing

FAV
 Fast Attack Vehicle
FBI
 Federal Bureau of Investigation
FBM
 Fleet Ballistic Missile [now SLBM]
FC
 (1) Field Circular (USA)
 (2) Fire Control
 (3) Fixed Cost
FCA
 Functional Configuration Audit
FCC
 Fleet Command Center (USN)
FCDNA
 Field Command Defense Nuclear Agency
FCDSSA
 Fleet Combat Direction System Support Activity [Dam Neck, Va.]
FCE
 Forward Command Element
FCLP
 Field Carrier Landing Practice (USN)
FCPE
 Naval Force Capabilities Planning Effort
FCRC
 Federal Contract Research Center
FCS
 (1) Facilities Checking Squadron
 (2) Fire Control System
FCT
 Foreign Comparative Test
FCTC
 Fleet Combat Training Center (USN)
FDC
 Fire Direction Center [ashore]
FDD
 Functional Description Document
FDDI
 Fiber Distributed Data Interface
FDESC
 Force Description

FDL
Fast Deployment Logistics
FDM
Frequency Division Multiplex
FDO
(1) Fighter Director Officer (USN)
(2) Fire Direction Officer
FDR
Final/Formal Design Review
FDR/FA
Flight Data Recorder/Fault Analyzer
FDS
Fixed Distribution System [acoustic ASW sensor]
FDTE
Force Development Test and Experimentation (USA)
FEBA
Forward Edge of Battle Area
FED
Forward Entry Device (USA)
FEDR
Full-scale Engineering Development Phase
FEMA
Federal Emergency Management Agency
FEML
Funded Environmental and Morale Leave Program
FEP
Fleet Satellite Communications Extremely High Frequency Package
FEPCA
Federal Employee Pay Comparability Act
FEU
40-foot container Equivalent Unit
FEWS
Follow-on Early Warning System [satellite]
FEWSG
Fleet Electronic Warfare Support Group (USN)
FFAR
Folding Fin Aerial Rocket
FFARP
Fleet Fighter Air [combat] Readiness Program (USN)
FFC-A
Forward Forces Command—Army

F³ (also FFF)
 Form-Fit-Function
FFP
 Firm Fixed-Price
FFRDC
 Federally Funded Research and Development Center
FFW
 Failure-Free Warranty
FHA
 Federal Housing Authority
FHE
 Forward Headquarters Element
FHTNC
 Fleet Hometown News Center [USN]
FIC
 (1) Fleet Intelligence Center
 (2) Flight Inspection Center
 (3) Force Indicator Code
FID
 Foreign Internal Defense
FIDP
 Foreign Internal Defense Plan
FIE
 Fly-In Echelon [Marines, supplies, and equipment deployed by strategic airlift during an operation] (USMC)
FISC
 Fleet and Industrial Supply Center [formerly Naval Supply Center, Norfolk, Va.; changed in 1993] (USN)
FISINT
 Foreign Instrumentation and Signals Intelligence
FISO
 Force Integration Staff Officer (USA)
FIST
 (1) Federal Information Processing Standards
 (2) Fire Support Team
 (3) Fleet Imagery Support Terminal
FIST-V
 Fire Integration Support Team Vehicle (USA)
FIT
 (1) Fault Isolation Tree [or Test]
 (2) Fleet Introduction Team (USN)

FITWEPSCOL
Fighter Weapons School [Topgun] (USN)
FJSRL
Frank J. Seiler Research Laboratory (USAF)
FL
Full Load
FLASH
Folding Light Acoustic Sonar for Helicopters
Fld
Field (USA)
FLDMEDSERVSCHOL
Field Medical Service School (USN)
FLEETEX
Fleet Exercise
FLIR
Forward-Looking Infrared [*not Radar*]
FLITE
Federal Legal Information Through Electronics
FLO/FLO
Flow-On/Flow-Off [very heavy lift ship]
FLOGEN
Flow Generator (USAF)
FLOT
(1) Flotilla [suffix; e.g. DESFLOT, CRUDESFLOT]
(2) Forward Line of Own Troops
FLT
Fleet
FLTCINC
Fleet Commander in Chief
FLTCORGRU
Fleet Coordinating Group
FLT CQ
Fleet Carrier Qualification
FLTMINWARTRACEN
Fleet Mine Warfare Training Center
FLTSAT
Fleet Satellite
FLTSATCOM
Fleet Satellite Communications [system]
FM
(1) Field Manual
(2) Financial Management

(3) Force Module
(4) Frequency Modulation
FMC
Fully Mission Capable
FMEA
Failure Mode and Effects Analysis
FMECA
Failure Mode and Effects Criticality Analysis
FMF
(1) Fleet Marine Force
(2) Foreign Military Financing
FMFIA
Federal Manager's Financial Integrity Act
FMFLANT
Fleet Marine Force Atlantic
FMFM
Fleet Marine Force Manual
FMFP
Foreign Military Financing Program
FMFPAC
Fleet Marine Force Pacific
FMI
Force Module Identifier
FML
Force Module Library
FMLP
Field Mirror Landing Practice (USN)
FMLSM
Force Module Logistics Sustainability Model
FMO
Frequency Management Office
FMOCC
Fleet Mobile Operations Command Center
FMOGDS
Field Medical Oxygen Generation/Distribution System
FMP
Fleet Modernization Program (USN)
FM&P
Force Management and Personnel
FMS
(1) Fleet Medical School
(2) Flexible Machining System

FMSCR

    (3) Force Module Subsystem
    (4) Foreign Military Sales

FMSCR

    Foreign Military Sales Credit

FMSS

    (1) Foreign Military Sales Financing
    (2) [See FLDMEDSERVSCHOL]

FMT

    Foreign Military Training

FMTV

    Family of Medium Tactical Vehicles (USA)

FN

    French Navy

FNOC

    Fleet Numerical Oceanographic Center

FO

    Forward Observer

FOA

    Field Operating Agency (USAF)

FOAS

    Field Operating Agencies (USAF)

FOB

    Forward Operating Base

FOC

    (1) First of Class
    (2) Full Operational Capability

FOD

    Foreign Object Damage [to aircraft engines]

FOFA

    Follow-On Forces Attack

FOG-M

    Fiber-Optical Guided Missile

FOI

    Freedom Of Information

FOIA

    Freedom Of Information Act

FOL

    Forward Operating Location

FON

    Freedom Of Navigation

FOR

    Force [suffix]

FORCAP
　　Force Combat Air Patrol (USN)
FORSCOM
　　Forces Command (USA)
FOSIC
　　Fleet Ocean Surveillance Center
FOSIF
　　Fleet Ocean Surveillance Facility
FOTC
　　Force Over-the-Horizon Target Coordinator
FOT&E
　　Follow-On Test and Evaluation
FOTRS
　　Follow-On Tactical Reconnaissance System (USAF)
FP
　　Firing Position (USA)
FPAF
　　Fixed Price Award Fee
FPBD
　　Functional Plan Block Diagram
FPCA
　　Federal Post Card Application [Post Card Registration and Absentee Ballot Request]
FPDS
　　Federal Procurement Data System
FP-E
　　Fixed-Price with Escalation
FPF
　　Final Protective Fire (USMC)
FPI
　　Fixed-Price Incentive
FPIF
　　Fixed-Price Incentive, Firm [target]
FPIS
　　Fixed-Price Incentive, Successive [target]
FPO
　　Fleet Post Office
FPT
　　Fleet Project Team
FQR
　　Formal Qualification Review

FRACAS
    Failure Reporting, Analysis, and Corrective Action System
Frag
    Fragmentation [weapon]
FRAG/FRAGO
    Fragmentary Order
FRAM
    Fleet Rehabilitation And Modernization (USN-USCG)
FRD
    Formerly Restricted Data [classification]
FREF
    Force Record Extract File
FRESSCAN
    Frequency Scan [radar]
FRG
    (1) Federal Republic of Germany
    (2) Force Requirement Generator
FRN
    Force Requirement Number
FROG
    Free Rocket Over Ground [Soviet—Russian]
FRP
    Full-Rate Production
FRS
    Fleet Readiness Squadron (USN)
FRWG
    Fleet Requirements Working Group
FS
    (1) Feasibility Study [Environmental Protection Agency]
    (2) Fire and Safety Technician (USCG)
    (3) Fire Support (USA)
FSA
    Fire Support Area
FSB
    Forward Support Battalion (USA)
FSC
    (1) File Server Control
    (2) Fire Support Center (USA)
    (3) Fire Support Coordinator (USMC)
FSCC
    Fire Support Coordination Center (USMC)

FSCL
　Fire Support Coordination Line (USA-USMC)
FSCM
　Federal Supply Code for Manufacturers
FSCOORD
　Fire Support Coordinator
FSCTT
　Fire Support Coordination Team Trainer
FSD
　Full-Scale Development [obsolete; now EMD]
FSED
　Full-Scale Engineering Development
FSG
　Federal Stock Group
FSM
　Firmware Support Manual
FSN
　Federal Stock Number
FSO
　Fire Support Officer (USA)
FSOCOM
　First Special Operations Command
FSPG
　Force Structure Planning Group (USMC)
FSS
　(1) Fast Sealift Ship (USN)
　(2) Federal Supply Schedule
FSSG
　Force Service Support Group (USMC)
FST
　Foreign Service Tour (USA)
FSU
　Fire and Safety Unit (USCG)
FSV
　Future Scout Vehicle (USA)
FSW
　Forward-Swept Wing
FTAM
　File Transfer, Access, and Management
FTC
　Force Track Coordinator

FTD
  (1) Foreign Technology Directorate
  (2) Foreign Technology Division (USAF)
FTEG
  Flight Test and Engineering Group (USN)
FTS
  (1) File Transfer Service
  (2) Full-Time Support
FTS2000
  Federal Telephone System 2000
FTTC
  Fleet Tactical Training Course
FTX
  Field Training Exercise (USA)
FUBAR
  F——d Up Beyond All Recognition [slang]
FUDS
  Formerly Used Defense Site
FUE
  First Unit Equipped
FUS
  Forward Support Unit
FWC/SWC
  Force/Ship Weapons Coordinator
Fwd
  Forward
FWW
  Follow-on Wild Weasel [aircraft] (USAF)
FY
  Fiscal Year
FYDP
  Future-Year Defense Program [formerly Five-Year Defense Plan; term adopted in 1991]

G

G1
Staff Officer for Personnel (USA-USMC)[4]
G2
Staff Officer for Intelligence (USA-USMC)
G3
Staff Officer for Operations (USA-USMC)
G4
Staff Officer for Supply/Logistics (USA-USMC)
G5
Staff Officer for Planning (USA-USMC)
GA
Tabun [nerve agent]
G&A
General and Administrative
G/AIT
Ground/Airborne Integrated Terminal (USAF)
GAO
General Accounting Office [*not* Government]
GAT
Government Acceptance Test
GATERS
Ground-Air Telerobotic Systems (USMC)
GATS
GPS-Aided Targeting System (USAF)
GB
Sarin [nerve agent]
GBD
Geometric Data Base
GBI
Ground-Based Interceptor (USA)
GBL
Government Bill of Lading
GBR
Ground-Based Radar (USA)

---

[4]The G-staff designations are used at senior staff levels; S-staff designations are used at lower echelons.

GBS
　Ground-Based Sensor
GBU
　Guided Bomb Unit
GCA
　Ground Controlled Approach
GCC
　Gulf Cooperation Council [Persian Gulf]
GCE
　Ground Combat Element (USMC)
GCI
　Ground Controlled Intercept
GCRES
　Ground Combat Readiness and Evaluation Squadron (USAF)
GD
　(1) General Dynamics Corporation
　(2) Soman [nerve agent]
G-Day
　First day of a ground campaign
GDIP
　General Defense Intelligence Program [fiscal]
GDLS
　General Dynamics Land Systems
GDP
　General Defense Plan
GE
　General Electric Company
GELOC/GEOLOC
　Standard Specific Geolocation Code
GENSER
　General Service [classification of message]
GEODSS
　Ground-Based Electrooptical Deep Space Surveillance (USAF)
GEOFILE
　Standard Specified Geographic Location File
GEOREF
　Geographic Reference System
GEOSAT
　Geodesy Satellite
GF
　nerve agent

GFAE
Government Furnished Aeronautical Equipment
GFC
Gunfire Control
GFCS
Gunfire Control System
GFE
Government Furnished Equipment
GFF
Government Furnished Facilities
GFI
Government Furnished Information
GFM
Government Furnished Material
GFOAR
Global Family of OPLANs Assessment Report
GFP
Government Furnished Property
GFS
Government Furnished Software
GHQ
General Headquarters
GHQAF
General Headquarters Air Force
GI
Government Issue [slang for U.S. soldier]
GIDEP
Government Industry Data Exchange Program
GIN
Greenland-Iceland-Norway [gap]
GIPSY
Graphic Information Presentation System
GIUK
Greenland-Iceland-United Kingdom [gap]
GKO
State Committee of Defense [Soviet]
GL
Great Lakes
GLBM
Ground-Launched Ballistic Missile
GLCM
Ground-Launched Cruise Missile [Tomahawk]

GLD
Ground Laser Designator
GLOBIXS
Global Information Exchange System
GLONASS
Global Navigation Satellite System [Soviet—Russian]
GM
Guided Missile [British]
GMF
Ground Mobile Force
GML
Guided Missile Launcher
GMR
Graduated Mobilization Response
GMS
Guided Missile System [British]
GMT
Greenwich Mean Time
GNP
Gross National Product
GOC
General Officer Commanding [British]
GOCO
Government-Owned, Contractor-Operated
GOGO
Government-Owned, Government-Operated
GOI
Government of Israel
GOR
Gradual-Onset-Rate (USAF)
GOSIP
Government Open Systems Interconnection Profile
GOTS
Government Off-The Shelf
GP
(1) General-Purpose
(2) Guided Projectile
GPALS
Global Protection Against Limited Strikes [part of SDI]
GPETE
General Purpose Electronic Test Equipment

GPH
    Gallons-Per-Hour
GPO
    Government Printing Office [Washington, D.C.]
GPS
    Global Positioning System [not satellite]
GPWS
    Ground Proximity Warning System (USAF)
GQ
    General Quarters
GRCA
    Ground Reference Coverage Area
GRCS
    Guard Rail Common Sensor (USA)
GRH
    Gramm-Rudman-Hollings [budget deficit control act]
GRP
    Glass-Reinforced Plastic
GRREG
    Graves Registration (USA)
GRT
    Gross Registered Tons
GRU
    (1) Chief Directorate of Intelligence of the General Staff [Soviet-Russian]
    (2) Group [suffix; e.g. PHIBGRU, SUBGRU] (USN)
GS
    (1) General Schedule
    (2) General Staff
    (3) General Support
GSA
    General Services Administration
GSBCA
    General Services Board of Contract Appeals
GSDF
    Ground Self-Defense Force [Japanese]
GSE
    Ground Support Equipment
GSFG
    Group of Soviet Forces [in] Germany
GSM
    Ground Station module (USA)

GSO
- (1) General Submarine Officer
- (2) Government Services Organization

GSR
Ground Surveillance Radar

GSTS
Ground-based Surveillance and Tracking System (USA)

GTMO
Guantanamo Bay, Cuba [pronounced GIT´-mo]

GTN
Global Transportation Network

GUPPY
Greater Underwater Propulsive Power [submarine modification] (USN)

G/VLLD
Ground/Vehicle Laser Locator and Designator (USA)

GWEN
Ground Wave Emergency Network

GWS
Guided Weapon System [British]

H

HA
- (1) Health Affairs
- (2) Holding Area (USA)

HAARP
High Frequency Active Auroral Research Program (USAF)

HAB
Heavy Assault Bridge (USA)

HAC
- (1) Helicopter Aircraft Commander (USN)
- (2) House Appropriations Committee

HAEMP
High-Altitude Electromagnetic Pulse

HAHO
  High-Altitude, High-Opening [parachute drop]
HAL
  Hindustan Aeronautics Limited [India]
HALO
  High-Altitude, Low-Opening [parachute jump]
HARDMAN
  Hardware/Manpower (USN-USMC)
HARM
  High-Speed Anti-Radiation Missile
HASC
  House Armed Services Committee
HAW
  Hypersonic Aerodynamic Weapon
HAZ
  Hazardous Cargo
HBC
  House Budget Committee
HC
  High Capacity [bombardment ammunition]
HCA
  (1) Head of Contracting Activity
  (2) Head of Contracting Agency
  (3) Humanitarian and Civic Assistance
HCO
  Helicopter Control Officer (USN)
HDC
  Helicopter Direction Center (USN)
HDTV
  High Definition Television
HDU
  Hose Down Unit
HE
  High Explosive
HEAT
  High-Explosive Anti-Tank
HEDI
  High Endo-Atmospheric Defense Interceptor
HEI
  High-Explosive Incendiary
Hel
  Helicopter

HEL
   High-Energy Laser
HELSTF
   High-Energy Laser System Test Facility (USA)
HEML
   High-Energy Microwave Laboratory [Kirtland AFB] (USAF)
HEMP
   High-Altitude Electro-Magnetic Pulse (USA)
HEMTT
   Heavy Expanded Mobility Tactical Truck
HERA
   High-Explosive Rocket Assisted (USA)
HERO
   Hazard of Electromagnetic Radiation to Ordnance (USN)
HET
   Heavy Equipment Transporter
HETS
   Heavy Equipment Transporter System (USA)
HF
   High Frequency
HFAJ
   High Frequency Anti-Jam [radio]
HFDF
   High Frequency Direction Finding
HF/DF
   High Frequency/Direction Finding ["huff/duff"]
HFE
   Human Factors Engineering
HFM
   Heavy Forces Modernization (USA)
HFMR
   HF Modem Replacement
HHG
   Household Goods
H-Hour
   Hour at which an operation or exercise begins
HHV
   Heavy High-mobility multipurpose wheeled Vehicle [see also HMMWV]
HIDACZ
   High Density Airspace Control Zone
HIFR
   Helicopter Inflight Refueling (USN)

HIGE
Hover In Ground Effect
HIMAD
High-to-Medium Range Air Defense (USA)
HIMARS
High Mobility Artillery Rocket System (USA)
HIP
Howitzer Improvement Program (USA)
HIRSS
Hover Infrared Suppressor Subsystem
HIRTA
High Intensity Radio Transmission Area (USA)
HIT
High Interest Track
HLA
Horizontal Line Array [sonobuoy]
HLG
High-Level Group
HLP
Heavy-Lift Preposition Ship
HLTF
High-Level Task Force
HLZ
Helicopter Landing Zone (USN)
HM
Hazardous Material
HMAS
Her Majesty's Australian Ship
HMCS
Her Majesty's Canadian Ship
HM&E
Hull, Machinery, and Electrical (USN)
HMG
Heavy Machine Gun
HMI
Human Machine Interface
HML
Hard Mobile Launcher (USAF)
HMMWV
High Mobility Multipurpose Wheeled Vehicle ["hum-vee"]
HMO
Health Maintenance Organization

HMS
　Her Majesty's Ship
HNS
　Host Nation Support
HNVS
　Helicopter Night Vision System
HOGE
　Hover Out-of-Ground Effect
HOI
　Headquarters Operating Instruction (USAF)
HOJ
　Home-On-Jam [missile]
HOL
　Higher Order Language
HOMS
　Hellfire Optimized Missile System (USA)
HONA
　Health of Naval Aviation
HOSTAC
　Helicopter Operations From Ships Other Than Aircraft Carriers Supplement
HOTAS
　Hands-On-Throttle-And-Stick (USAF-USN)
How
　Howitzer
hp
　horsepower
HPI
　High-Power Illuminator [radar] (USN)
HPT
　Horsepower Tonnage
HQ
　Headquarters
HQDA
　Headquarters, Department of the Army
HQMC
　Headquarters Marine Corps (USMC)
HRO
　Housing Referral Office, Germany
HRT
　Hostage Rescue Team [FBI]

HSB
: High Speed Boat

HSC
: Health Services Command (USA)

HSD
: Human Systems Division (USAF)

HSDB
: High Speed Data Bus

HSEP
: Hospital Surgical Expansion Package (USAF)

HSFB
: High Speed Fleet Broadcast

HSS
: Health Service Support (USA)

HSU
: Hero of the Soviet Union [award]

HTK
: Hard-Target Kill

HTKP
: Hard-Target Kill Potential

HTP
: High-Test Peroxide

HTS
: High-Tensile Steel

HTSSE
: High-Temperature Super-Conductivity Space Experiment (USN)

HTTB
: High-Technology Test Bed [C-130 advanced technology flight demonstrator]

HTV
: Hull Test Vehicle

HU
: Hospital Unit

HUD
: Head-Up Display [singular, *not* Heads]

HUK
: Hunter Killer [ASW; no longer used] (USN)

HUMINT
: Human Intelligence

HVAA
: High-Value Airborne Assets

HVT
> High-Value Target

HVU
> High-Value Unit

HVUCAP
> High-Value Unit Combat Air Patrol (USN)

HW
> (1) Hardware
> (2) Hazardous Waste

H/W
> Hardware

HWIC
> Hardware Configuration Item

HWR
> Heavy Water Reactor

HWSTD
> High Water Speed Technology Demonstrator (USMC)

HWT
> Heavy-Weight Torpedo

HY
> High Yield [material strength]

I

I²
> Image Intensification

IA
> Intelligence Assessment

IADC
> Inter-American Defense College

IADS
> Integrated Air Defense System

IAEA
> (1) International Atomic Engergy Accord
> (2) International Atomic Energy Agency

IAF
　Israel Air Force
IAI
　Israel Aircraft Industries, Limited
IAMP
　Imagery Acquisition and Management Plan
IAS
　Indicated Airspeed (USN)
IATACS
　Improved Army Tactical Communications System
IB
　Issue Book
IBAF
　Interim Brigade Afloat Force [prepositioning force] (USA)
IBAHRS
　Inflatable Body and Head Restraint System
IBP
　Industrial Base Program
IBPDMS
　Improved Point Defense Missile System [Sea Sparrow]
IC
　Interior Communications
ICA
　(1) Independent Cost Analysis
　(2) Integrated Communications Architecture (USN)
ICAAS
　Integrated Control and Avionics for Air Superiority (USAF)
ICADS
　Integrated Correlation and Display System (USAF)
ICAF
　Industrial College of the Armed Forces
ICAO
　International Civil Aviation Organization
ICAP
　Improved Capabilities
ICB
　Imitative Communication Deception
ICBM
　Intercontinental Ballistic Missile
ICD
　Interface Connecting Device (USA)

ICDS
Improved Conventional Dive System
ICE
Independent Cost Estimate
ICG
Interactive Computerized Graphics
ICLS
Instrument Carrier Landing System (USN)
ICM
Improved Conventional Munitions (USA)
ICMMP
Integrated CONUS Medical Mobilization Plan
ICNIA
Integrated Communication, Navigation, Identification Avionics (USAF)
ICOD
Intelligence Cutoff Date
ICON
(1) Imagery Communications and Operations Node
(2) Integrated COMSEC (USA)
ICP
Inventory Control Point
ICR
(1) Intelligence Collection Requirement
(2) Intercooled Recuperative [engine]
ICRP
Internal Control Review Program (USAF)
ICRS
International Committee of the Red Cross/Crescent
ICS
(1) Intelligence Center and School (USA)
(2) Intelligence Community Staff
(3) Intercockpit Communications System (USN)
(4) Interim Contractor Support
(5) Intercommunication System (USN)
ICV
Infantry Combat Vehicle [generally replaced in usage by IFV]
ICWG
Interface Control Working Group
ID
(1) Identification
(2) Identifier
(3) Increased Deployability posture

IDA
  Institute for Defense Analyses
IDAP
  Integrated Defensive Avionics Program (USN)
IDAS
  Integrated Defense Avionics System (USAF)
IDD
  Interface Design Document
IDF
  Israel Defense Forces
IDHS
  Intelligence Data Handling System
IDM
  Improved Data Modem (USAF)
IDS
  Interdictor/Strike
IE
  Industrial Engineer
IED
  Integrated Electric Drive (USN)
IES
  (1) Illustrative Evaluation Scenario
  (2) Imagery Exploitation System
  (3) Industrial Engineering Standard
IEW
  Intelligence and Electronic Warfare (USA)
IF
  Industrial Fund
IFB
  Invitation For Bid
IFC
  Instrument Flight Center (USAF)
IFF
  Identification, Friend or Foe
IFR
  Instrument Flight Rules
IFSAS
  Interim Fire Support Automation System (USA)
IFT
  Industrial Field Trip
IFTE
  Integrated Family of Test Equipment (USA)

IFV
    Infantry Fighting Vehicle
IG
    (1) Inspector General
    (2) Interdepartment Group (JCS)
IGCE
    Independent Government Cost Estimate
IGE
    In-Ground Effect [hover]
IGRV
    Improved Guard Rail V (USA)
IGSM
    Interim Ground Station Module [JSTARS]
IGV
    Inlet Guide Vanes (USN)
IHADSS
    Integrated Helmet and Display Sight System (USA)
IHE
    Insensitive High Explosive
IHPTET
    Integrated High-Performance Turbine Engine Technology (USAF)
I&I
    Inspector and Instructor [for reserve units] (USMC)
IIP
    Initial Issue Provisioning (USMC)
IIR
    (1) Imaging Infrared
    (2) Intelligence Information Report
IISS
    International Institute for Strategic Studies [London]
IKPT
    Initial Key Personnel Training (USA)
Il
    Ilyushin [Soviet-Russian aircraft designation]
I&L
    Installation and Logistics (USMC)
ILC
    (1) Improved Line Charge
    (2) International Logistics Center (USAF)
ILCOS
    Instantaneous Lead Computing Optical Sight [gunsight] (USN)

ILS
- (1) Instrument Landing System (USN)
- (2) Integrated Logistics Support

ILSMT
ILS Management Team

ILSP
ILS Plan

ILW
Institute of Land Warfare [Association of the U.S. Army]

IM
Item Manager

IMA
- (1) Individual Mobilization Augmentees [or Augmentation] (reservist)
- (2) Intermediate Maintenance Activity (USN)

IMC
- (1) Instrument Meteorological Condition (USN)
- (2) Internal Management Control

IMCO
International Maritime Consultive Organization

IMET
International Military Education and Training

IMINT
Imagery Intelligence

IMIP
- (1) Industrial Modernization Improvement Program (USAF)
- (2) Industrial Modernization Incentives Program

IML
Intermediate Maintenance Level

IMOM
Improved Many On Many (USAF)

IMP
Internal Management Control

IMU
Inertial Measurement Unit

INC
Insertable Nuclear Component

INCA
Intelligence Communications Architecture

INCNR
Increment Number

INEWS
   Integrated Electronic Warfare System
Inf
   Infantry
INF
   Intermediate-Range Nuclear Forces [treaty]
INFIL/EXFIL
   Infiltration and Exfiltration
ING
   Inactive National Guard
INMARSAT
   International Maritime Satellite
INNF
   Intermediate Naval Nuclear Forces
INRC
   Innovative Naval Reserve Concept
INS
   Inertial Navigation System
INSCOM
   (1) Integrated Special Intelligence Communications Architecture
   (2) Intelligence and Security Command (USA)
INSMARSAT
   International Maritime Satellite [Organization]
Intel
   Intelligence
INTELCAST
   Intelligence Broadcast
INTELNET
   Intelligence Network
IO
   (1) Indian Ocean
   (2) Intelligence Officer
   (3) Intelligence Oversight
IOBC
   Infantry Officer Basic Course (USA)
IOC
   (1) Initial Operational Capability
   (2) Intelligence Operations Center (USAF)
IOM
   International Organization for Migration
IOT&E
   Initial Operational Test and Evaluation (USA)

IP
- (1) Industry Program [DSMC]
- (2) Initial Point
- (3) Initial Production
- (4) Intelligence Policy

IPB
- (1) Intelligence Preparation of Battlefield (USA-USMC)
- (2) Intelligence Preparatory Brief (USA)

IPCE
Independent Parametric Cost Estimate

IPDS
- (1) Imagery Processing and Dissemination System
- (2) Improved Point Defense Missile System (USN)

IPE
- (1) Increased Performance Engine (USAF)
- (2) Industrial Plant Equipment

IPF
Initial Production Facilities

IPL
Integrated Priority List

IPP
Industrial Preparedness Planning

IPR
- (1) In-Process Review
- (2) In-Progress Review

IPS
- (1) Illustrative Planning Scenario (USA)
- (2) Imagery Processing System (USA)
- (3) Integrated Program Summary

IPSS
Initial Pre-planned Supply Support

IPT
Intermediate Phase Training

IR
- (1) Information Requirements (USA)
- (2) Infrared
- (3) Intelligence Request

IRA
Interim Response Actions (USA)

IRAD
Industry independent Research and Development

IRADS
Infrared Acquisition and Designation System

IRAN
    Inspection and Repair As Necessary (USN-USAF)
IRBM
    Intermediate-Range Ballistic Missile [now INF or LRINF missile]
IRD
    Infrared Detection Set
IR&D
    Independent Research and Development
IRDS
    Infrared Detection Set
IRFMS
    Interservice Radio Frequency Management School
IRM
    Information Resources Management (DOD)
IRP
    Installation Restoration Program
IRR
    Individual Ready Reserve
IRS
    Interface Requirements Specification
IRST
    Infrared Search and Track [system]
ISA
    (1) Instruction Set
    (2) Intermediate Supply Activity (USMC)
    (3) International Security Affairs (DOD)
I-S/A AMPE
    Inter-Service Agency Automated Message Processing Exchange
ISAR
    Inverse Synthetic Aperture Radar (USN)
ISB
    Initial Staging Base (USA)
ISD
    Instructional Systems Development (USAF)
ISDN
    Integrated Services Digital Network
ISE
    (1) Independent Ship Exercise (USN)
    (2) Intelligence Support Element
ISEA
    In-Service Engineering Agent (USN)

IS-IS
　Intermediate System-to-Intermediate System
ISM
　Industrial Security Manual
ISO
　International Standards Organization
ISP
　Integrated Support Plan
IS&T
　Innovative Science and Technology
ISU
　Integrated Sight Unit (USA)
ITA
　Intermediate Training Assessment
ITAC
　(1) Integrated Tactical Aircraft Control (USAF)
　(2) Intelligence and Threat Analysis Center (USA)
ITALD
　Improved Tactical Air Launched Decoy (USN)
ITAR
　International Traffic in Arms Regulations
ITDN
　Integrated Tactical-Strategic Data Network
ITF
　Intelligence Task Force
ITO
　Installation Transportation Office (USA)
ITP
　Integrated Test Plan
ITS
　Integrated Training System
ITV
　Improved TOW Vehicle (USA)
IUS
　Inertial Upper Stage
IUSS
　Integrated Undersea Surveillance System (USN)
IV&V
　Independent Verification and Validation
I&W
　Indications and Warning

IWSM
　Integrated Weapon Systems Management (USAF)
IWSO
　Instructor Weapon Systems Officer (USAF)

J

J1
　(1) Joint Staff Officer for Personnel
　(2) Manpower and Personnel Directorate (JCS)
J2
　(1) Joint Staff Officer for Intelligence
　(2) Defense Intelligence Agency (JCS)
J3
　(1) Joint Staff Officer for Operations
　(2) Operations Directorate (JCS)
J4
　(1) Joint Staff Officer for Logistics
　(2) Logistics Directorate (JCS)
J5
　(1) Joint Staff Officer Plans
　(2) Strategic Plans and Policy Directorate (JCS)
J6
　(1) Command, Control, and Communications Systems Directorate (JCS)
　(2) Joint Staff Officer for Communications
J7
　(1) Joint Staff Officer for Operational Plans and Interoperability
　(2) Operational Plans and Interoperability Directorate (JCS)
J8
　Force Structure, Resource, and Assessment Directorate (JCS)
J&A
　Justification and Approval
JAAT
　Joint Air Attack Team (USA)

JAC
Job Assistance Center (USA)
JACADS
Johnston Atoll Chemical Agent Disposal System (USA)
JACC
Joint Air Command Center (USA)
JADO
Joint Air Defense Operations (USMC)
JAF
Joint Attack Fighter (USAF-USN-DOD)
JAG
Judge Advocate General
JAIC
Joint Air Intelligence Center
JAMAC
Joint Aeronautical Materials Activity
JANAP
Joint Army, Navy, Air Force Publication
JAO
Joint Area of Operations
JASDF
Japanese Air Self-Defense Force
JASORS
Joint Advanced Special Operations Radio System
JAST
Joint Advanced Strike Technology [program] (USAF-USN)
JATO
Jet Assisted Take-Off [*rocket* booster]
JBD
Jet Blast Deflector (USN)
JBPO
Joint Blood Program Office
JC^{3}CMSOC
Joint C^{3}CM Staff Officer Course
JC^{3}CM STB
Joint C^{3}CM Senior Tactical Battle Commanders Course
JCACC
Joint Combat Airspace Command and Control Course
JCALS
Joint CALS
JCASS
Joint Contract Administration Coordinating Council

JCC
  Joint Coordination Center
JCEOI
  Joint Communications Electronics Operations Instructions
JCGRO
  Joint Central Graves Registration Office
JCLL
  Joint Center for Lessons Learned
JCMC
  Joint Crisis Management Capability
JCS
  Joint Chiefs of Staff (DOD)
JCSE
  Joint Communications Support Element
JCSM
  Joint Chiefs of Staff Memorandum
JDA
  (1) Japan Defense Agency
  (2) Joint Deployment Agency
  (3) Joint Duty Assignment
JDAL
  Joint Duty Assignment List
JDAM
  Joint Direct Attack Munition
JDAMIS
  Joint Duty Assignment Management Information System
JDAP
  Joint Direct Attack Program (USAF)
JDS
  Joint Deployment System
JDSSC
  Joint Data Systems Support Center
JEWSOC
  Joint Electronic Warfare Staff Officer Course
JEZ
  Joint Engagement Zone (USMC)
JFACC
  Joint Force Air Component Commander
JFC
  (1) Joint Force Commander
  (2) Joint Forces Command

JFDP
  Joint Force Development Process
JFLC
  Joint Forces Land Component
JFLCC
  Joint Forces Land Component Commander
JFTR
  Joint Federal Travel Regulations
JIAD
  Joint Integrated Avionics Directorate
JIAWG
  Joint Integrated Avionics Working Group
JIC
  Joint Intelligence Center
JICPAC
  Joint Intelligence Center Pacific
JIF
  Joint Interrogation Facility
JIIKS
  Joint Imagery Interpretation Keys Structure [Defense Intelligence Agency]
JILE
  Joint Intelligence Liaison Element
JINTCCS
  Joint Interoperability of Tactical Command and Control Systems
JIPC
  Joint Imagery Production Complex
JIT
  Just-In-Time
JITC
  Joint Interoperability Center
JLC
  Joint Logistics Commanders
JLPPG
  Joint Logistics and Personnel Policy and Guidance
JLRSS
  Joint Long-Range Strategic Study
JLSC
  Joint Logistics System Command
JMA
  Joint Mission Analysis

JMEMS
  Joint Munition Effectiveness Manual (USN)
JMETL
  Joint Mission Essential Task List
JMNA
  Joint Military Net Assessment [JCS] (DOD)
JMPAB
  Joint Material Priorities and Allocation Board
JMRO
  Joint Medical Regulating Office
JMSDF
  Japan Maritime Self-Defense Force
JMSNS
  Justification for Major System New Start
JO
  (1) Journalist [slang] (USN)
  (2) Junior Officer [slang] (USN)
JOA
  Joint Operating Agreement
JOC
  Joint Operations Center
JOOD
  Junior Officer Of the Deck (USN)
JOP
  Joint Operating Procedures
JOPES
  Joint Operations, Planning, and Execution System
JOPS
  Joint Operation Planning System
JOPSREP
  JOPS Reporting System
JOTC
  Jungle Operations Training Center
JOTS
  Joint Operational Tactical System (USN)
JP-4/5
  Aviation jet fuel
JPAO
  Joint Public Affairs Office
JPATS
  Joint Primary Aircraft Training System (USAF-USN)

JPEC
    Joint Planning and Execution Community
JPI
    Joint Precision Interdiction (NATO)
JPL
    Jet Propulsion Laboratory [California Institute of Technology, Pasadena]
JPME
    Joint Professional Military Education
JPO
    (1) Joint Petroleum Office
    (2) Joint Program Officer
JPP
    Joint Planning Process
JPTS
    Jet Petroleum, Thermally Stable
JQ1
    Chief of Staff, Joint Staff
JRC
    Joint Reconnaissance Center
JRCC
    Joint Rescue Coordination Center
JRMB
    Joint Requirements and Management Board
JROC
    Joint Requirements Oversight Council
JRS
    Joint Reporting Structure
JRSC
    Jam-Resistant Secure Communications
JRTC
    Joint Readiness Training Center [Fort Chaffee, Ark.]
JS
    Joint Staff
JSCAT
    Joint Staff Crisis Action Team
JSCC
    Joint Service Coordination Committee
JSCP
    Joint Strategic Capabilities Plan [JSPS]
JSDF
    Japanese Self-Defense Force

JSEAD
  Joint Suppression of Enemy Air Defense
JSIC
  Joint Space Command Intelligence Center
JSIPS
  Joint Service Imagery Processing System
JSO
  Joint Specialty Officer
JSOC
  (1) Joint Special Operations Center
  (2) Joint Special Operations Command
JSONOM
  Joint Specialty Officer Nominee
JSOR
  (1) Joint Service Operational Requirement
  (2) Joint Statements Of Requirements
JSORD
  Joint System Operational Requirements Document
JSOTF
  Joint Special Operations Task Force
JSOW
  Joint Stand-off Weapon (USAF-USN)
JSPACE
  Joint Space Intelligence/Operations Course
JSPACEINT
  Joint Space Intelligence/Operations Senior Course
JSPS
  Joint Strategic Planning System
JSR
  Joint Strategic Review
JSS
  Joint Surveillance System
JSSA
  Joint Stealth Strike Aircraft (DOD)
JSTARS
  Joint Surveillance/Target Attack Radar System
JSTPS
  Joint Strategic Target Planning Staff
JTAAMO
  Joint Tactical Air-to-Air Office
JTB
  Joint Transportation Board

JTC³A
 Joint Tactical Command, Control, and Communications Agency [merged with DCA]
JTDE
 Joint Technology Demonstrator Engine
JT&E
 Joint Test and Evaluation
JTF
 Joint Task Force
JTFME
 Joint Task Force Middle East
JTFP
 Joint Tactical Fusion Program
JTIDS
 Joint Tactical Information Distribution System
JTO
 JOPES Training Organization
JTR
 Joint Travel Regulations
JTTP
 Joint Tactics, Techniques, and Procedures
JULLS
 Joint Universal Lessons Learned System
JUMPS
 Joint Uniform Military Pay System
JUSMAG
 Joint U.S. Military Advisor Group
JUSMAG-K
 Joint U.S. Military Advisor Group—Korea
JUSMMAT
 Joint U.S. Military Mission for Aid to Turkey
JUWTF
 Joint Unconventional Warfare Task Force
JVIDS
 Joint Visually Integrated Display System
JVIS
 Joint Visual Information Services (DOD)
JVX
 Joint tilt-rotor aircraft project [subsequently designated V-22 series]
JWOD
 Javits-Wagner-O'Day [legislation setting requirements of DOD to use disadvantaged businesses]

K

Ka
　Kamov [Soviet-Russian aircraft designation]
KAF
　Kuwaiti Air Force
KATUSA
　Korean Augmentees to the U.S. Army
KE
　Kinetic Energy [weapon]
KEM
　Kinetic Energy Missile (USA)
KEW
　Kinetic Energy Weapons
KGB
　Committee for State Security [Soviet]
kgst
　kilograms static thrust
KH
　Key Hole [reconnaissance satellite series]
KIA
　Killed In Action
KIAS
　Knots Indicated Air Speed
KIDS
　Knowledge-Based Integrated Design System
KIO
　Kick It Off [slang]
KISS
　(1) Keep It Simple Stupid [slang]
　(2) Korean Intelligence Support System
KITE
　Kinetic kill vehicle Integrated Technology Experiment (USA)
KKMC
　King Khalid Military City [Saudi Arabia]
KKV
　Kinetic Kill Vehicle
KO
　Contracting Officer

KR
 Contractor
KT
 Contract
KTF
 Kuwaiti Civil Affairs Task Force
KTO
 Kuwaiti Theater of Operations
KUSLO
 Kenya-U.S. Liaison Office
KY
 (1) Keyhole [U.S. reconnaissance satellite]
 (2) U.S. prefix for Soviet-Russian developmental missiles [for Kapustin Yar test facility]

L

LA
 (1) Legislative Affairs
 (2) Legislative Assistant
LAAD
 Low-Altitude Air Defense (USMC)
LAAM
 Light Anti-Aircraft Missile (USMC)
LAAPS
 Laptop Automated Aid Positioning System (USCG)
LAB
 Light Assault Bridge (USA)
LABCOM
 Laboratory Command (USA)
LABS
 Low-Altitude Bombing System
LAD
 Latest Arrival Date [at POD]
Ladar**
 Laser radar

LAI
Light Armored Infantry (USMC)
LAMP-H
Lighter, Amphibian, Heavy Lift (USA)
LAMPS
Light Airborne Multipurpose System [helicopter] (USN)
LA/MPSS
Large Area/Mobile Projected Smoke System (USA)
LAN
Local Area Network
LANA
Low-Altitude Night Attack
LANL
Los Alamos National Laboratory [New Mexico]
Lant
Atlantic
LANTCOM
Atlantic Command (see also USACOM)
LANTIRN
Low-Altitude Navigation and Targeting Infrared for Night [system] (USAF)
LAO
Logistics Assistance Officer (USA)
LAPES
Low-Altitude Parachute Extraction System
LAR
Logistics Assistance Representative (USA)
LARA
Light Armed Reconnaissance Airplane
laser**
light amplification by simulated emission of radiation
LASH
Lighter Aboard Ship
LASS
Large Area Screening System (USA)
LAT
Low-Altitude Tactics
LAU
Launcher Armament Unit (USN)
LAURA
Low-Altitude Unmanned Reconnaissance Aircraft

LAV
  Light Armored Vehicle (USMC)
LAV-AD
  Light Armored Vehicle—Air Defense (USMC)
LAV-AG
  Light Armored Vehicle—Assault Gun (USMC)
LAVB
  Light Armored Vehicle Battalion (USMC)
LAW
  Light Anti-tank Weapon
LAWS
  Light Attack Weapons School (USN)
lbst
  pounds static thrust
LBT
  Land-Based Tanker [aircraft]
LBTS
  Land-Based Test Site
LCAC
  Landing Craft Air Cushioned
LCC
  (1) Land Component Commander (USA)
  (2) Launch Control Center (USAF)
  (3) Life Cycle Cost
  (4) Limited Capability Configuration (USA)
LCE
  Logistics Capability Estimator
LCM
  Life Cycle Management
LCN
  Load Classification Number
LCS
  (1) Low-Cost Seeker
  (2) Low-Cost Sonobuoy
LD
  (1) Line of Departure
  (2) Loaded Deployability posture
LDB
  Low Drag Bomb
LDC
  Less-Developed Country

LDO
   Limited Duty Officer (USN)
LDP
   Logistics Development Program
LEA
   Law Enforcement Agencies
LEAP
   Light Exo-Atmospheric Projectile [formerly *Lightweight*]
LEASAT
   Leased Satellite Communications System
LEC
   LAMPS Element Coordinator (USN)
LED
   Law Enforcement Detachment (USCG)
LEDET
   Law Enforcement Detachment (USCG)
LEIP
   Link Eleven Improvement Program
LEM
   Logistic Element Manager
LERTCON
   Alert Condition
LES
   Leaving and Earning Statement
LET
   Light Equipment Transporter
LF
   Low Frequency
LFA
   Low Frequency Active
LFAS
   Low-Frequency Active Sonar
LFF
   Logistic Factors File
LFTC
   Landing Force Training Command
LFT&E
   Live Fire Test and Evaluation
LGB
   Laser-Guided Bomb
LGW
   Laser-Guided Weapon

LH
  Light Helicopter (USA)
LHT
  Line-Haul Tractor
LHTEC
  Light Helicopter Turbine Engine Company [firm developing the power plant for the RAH-66 Comanche]
LHX
  Light Helicopter Experimental (USA)
LIBMISH
  U.S. Military Mission, Liberia
LIC
  Lower-Intensity Conflict [previously Low-Intensity Conflict]
LID
  Light Infantry Division (USA)
LIDAR
  High-resolution airborne electro-optical imaging device
LIFT
  Lead-In Flight Training (USAF)
LIT
  Lead-In Training (USAF)
LL
  Legislative Liaison
LLFA
  Low–Low Frequency Acoustics
LLLGB
  Low-Level Laser-Guided Bomb
LLLTV
  Low-Light-Level Television
LLOC
  Land Line of Communication
LLT
  Long Lead Time (USN)
LLTM
  Long Lead-Time Material
LM
  Logistics Management
LMDC
  Leadership and Management Development Center (USAF)
LMET
  Leadership Management and Education Training School

LMI
  Logistics Management Institute (DOD)
LMTV
  Light Medium Tactical Vehicle
LNG
  Liquified Natural Gas
LNO
  Liaison Officer
LO
  Low Observable
LOA
  (1) Length Overall [ship]
  (2) Letter Of Authorization
  (3) Letter of Offer and Acceptance
  (4) Limit of Advance (USA)
LOAL
  Lock-On After Launch [missile]
LOBL
  Lock-On Before Launch [missile]
LOC
  (1) Linked Operational Capability
  (2) Logistics Operations Center (USAF)
LOC(s)
  Line(s) Of Communication
LOE
  (1) Letter Of Evaluation (USAF)
  (2) Level Of Effort
  (3) Light-Off Examination (USN)
LOFAR
  Low-Frequency Acquisition and Ranging
LOFARGRAM
  Low-Frequency Analyzing and Recording Gram
LOG
  Logistics
LOGAIR
  Logistics Airlift (USAF)
LOGAIS
  Logistics Automated Information System (USMC)
LOGAMP
  Logistics and Acquisition Management Program (USA)

LOGCAP
(1) Logistics Civil Augmentation Program
(2) Logistics Command Assessment of Projects

LOGEX
Logistics Exercise (USA)

LOGMARS
Logistical Application of Automated Marking and Reading Symbology

LOGO
Limitation Of Government Obligation

LOGREG
Logistic Requirement Report (USN)

LOGSAFE
Logistics Sustainability Analysis Feasibility Estimator

LOI
(1) Letter Of Instruction
(2) Letter Of Intent

LOR/A
Level Of Repair/Analysis

LORAN
Long-Range Aid to Navigation

LOROP
Long-Range Oblique Photography

LOS
(1) Law Of the Sea
(2) Line Of Sight

LOSAT
Line-Of-Sight Anti-Tank

LOS-F
Line-of-Sight Forward Element

LOS-F-H
Line-of-Sight-Forward-Heavy (USA)

LOS-R
Line-of-Sight-Rear (USA)

LOTS
Logistics Over The Shore

LOX
Liquid Oxygen

LP
Limited Procurement

LPAR
Large Phased-Array Radar

LPI
　　Low Probability of Intercept
LPU
　　(1) Life Preserver Unit (USN)
　　(2) Limited Procurement Urgent
LPV
　　Laser-Protective Visor
LRA
　　Long-Range Aviation [Soviet; no longer used]
LRAACA
　　Long-Range Air ASW Capable Aircraft [P-3 Orion replacement program, subsequently designated P-7]
LRC
　　Logistics Readiness Center (USAF)
LRCCM
　　Long-Range Conventional Cruise Missile (USAF)
LRCSW
　　Long-Range Conventional Standoff Weapon
LRD
　　Laser Range Designator (USA)
LRE
　　Latest Revised Estimate
LRF
　　Laser Range Finder (USA)
LRF/D
　　Laser Range Finder/Designator (USA)
LRG
　　Logistics Review Group (USN)
LRI
　　Long-Range International
LRINF
　　Longer-Range Intermediate-Range Nuclear Forces
LRIP
　　Low-Rate Initial Production
LRMP
　　(1) Legacy Resource Management Program
　　(2) Long-Range Maritime Patrol [aircraft]
LRP
　　Low Rate Production
LRR
　　Long-Range Radar

LRRDAP
  Long-Range Research Development and Acquisition Plan (USA)
LRSI
  Long-Range SOF Insertion
LRU
  Line-Replaceable Unit (USAF)
LRU/LRM
  Line-Replaceable Unit/Module
LSA
  Logistics Support Analysis
LSAR
  Logistics Support Analysis Record (USA)
LSC
  Large-Scale Computer (USN)
LSE
  Landing Signal Enlisted (USN)
LSI
  Large-Scale Integration
LSO
  Landing Signal Officer (USN)
LST
  Laser Spot Tracker
LSV
  (1) Large-Scale Vehicle (USN)
  (2) Logistics Support Vessel (USA)
LTA
  Lighter-Than-Air [airships, blimps]
LTBT
  Limited Test Ban Treaty [U.S.-USSR, 1973]
LTD
  Laser Target Designator
LTD/R
  Laser Target Designator/Ranger [pod]
LTE
  Loss-of-Tail Rotor Effect
LTOE
  L-Series Tables of Organization and Equipment
LTSDE
  Low-Temperature Superconducting Device Electronics
LTV
  Ling-Temco-Vought [firm; now Vought]

LUT
Limited User Test
LVS
Logistics Vehicle System (USMC)
LWCSS
Light Weight Camouflage Screening System (USA)
LWIR
Long-Wave Infrared
LWT
Lightweight Torpedo [now Mk 50]
LZ
Landing Zone [helicopter]

M²C²
Multi-Media Communication Control
MAA
(1) Master At Arms
(2) Mission Area Analysis (USMC)
(3) Mission Area Assessment
MAAC
Military Assistance Advisory Command
MAAG
Military Assistance Advisory Group
MAARP
Medium Attack Advanced Readiness Program (USN)
MAB
Marine Amphibious Brigade [changed to MEB in 1988] (USMC)
MAC
Military Airlift Command [obsoleted, deactivated in June 1992] (USAF)
MACC
Modified Air Control Center (USAF)
MACCS
Marine Corps Air Command and Control System (USMC)

MACE
 Military Airlift Capability Estimator
MACOM
 Major Command (USA)
MACS
 Magnetic Countermine System
MACSAT
 Multiple Access Commercial Satellite
MAD
 (1) Magnetic Anomaly Detection [Detector]
 (2) Mission Area Deficiency
 (3) Mutual Assured Destruction
MADAR
 Malfunction Detection Analysis and Recording
MADOM
 Magnetic Acoustic Detection of Mines
MAF
 Marine Amphibious Force [changed to MEF in 1988] (USMC)
MAFC
 MAGTF All-Source Fusion Center
MAG
 (1) Maritime Action Group [non-carrier naval task group] (USN)
 (2) Maritime Aircraft Group
 (3) Military Advisory Group
MAGIS
 Marine Air Ground Intelligence System
MAGTF
 Marine Air-Ground Task Force (USMC)
Maint
 Maintenance
MAISARC
 Major Automated Information System Review Council
MAJCOM
 Major Command (USAF)
MALT
 Monetary Allowance in Lieu of Transportation
MAMP
 Mission Area Matériel Plan (USA)
MAN
 Metropolitan Area Network

MANPRINT
　Manpower and Personnel Integration Program (USA)
MANTECH
　Manufacturing Technology
MAP
　(1) Master Attack Plan
　(2) Military Assistance Program
MAPP
　Modern Aids to Planning Program
MAPS
　(1) Mobile Aerial Port Squadron (USAF)
　(2) Mobility Analysis and Planning System [MTMC]
MAR
　(1) Major Aircraft Review
　(2) Management Assessment Review (USAF)
　(3) Marine
MARAD
　Maritime Administration
MARC
　Manpower Requirement Criteria
MARCENT
　Marine Corps, U.S. Central Command [see USMARCENT]
MARCORPS
　U.S. Marine Corps
MARCORSYSCOM
　Marine Corps Systems Command
MARDEZ
　Maritime Defense Zone (USN-USCG)
MARDIV
　Marine Division
MARG
　Marine Amphibious Ready Group
MARLO
　Marine Liaison Officer
MARS
　Military Affiliated Radio Station
MARSAT
　Maritime Satellite
MaRV
　Maneuvering Reentry Vehicle
MARV
　Mobile Armored Reconnaissance Vehicle

MASH
Mobile Army Surgical Hospital
MASINT
Measurement And Signature Intelligence
MASS
MICAP [Mission Critical Parts] Asset Sourcing System
MAST
Military Assistance to Safety and Traffic
MATDEV
Material Developer (USA)
MATE
Multipurpose Automatic Test Equipment
MATS
Military Air Transport Service [now MAC] (USAF)
MATSG
Marine Aviation Training Support Group
MAU
(1) Marine Amphibious Unit [changed to MEU in 1988] (USMC)
(2) Master Augmentation Unit (USN)
MAWS
Missile Approach Warning System
MBA
Main Battle Area (USA)
MBFR
Mutual and Balanced Force Reductions
M-box
Maneuver box [subdivision of a kill box] (USMC)
MBPO
Military Blood Program Office
MBT
Main Battle Tank
MC
(1) Medical Corps (USA-USN)
(2) Military Committee (NATO)
(3) Military Construction
(4) Mission Capable
MCA
(1) Marine Corps Association
(2) Military Civic Action
(3) Military Construction, Army
MCAF
Marine Corps Air Facility

MCAGCC
 Marine Corps Air-Ground Combat Center [Twenty-nine Palms, Calif.]
MCAS
 Marine Corps Air Station
MCC
 Military Coordinating Committee
MCCDC
 Marine Corps Combat Development Command
MCCP
 Marine Corps Capabilities Plan
MCCR
 Mission Critical Computer Resources
MCCRES
 Marine Corps Combat Readiness Evaluation System
MCCS
 Mission Critical Computer System
MCE
 Modular Control Equipment
MCG
 Marine Corps Gazette [magazine]
MCIC
 Marine Corps Intelligence Center
MCLB
 Marine Corps Logistics Base
MCLWG
 Major Caliber Lightweight Gun [8-inch MK 71; cancelled project]
MCM
 (1) Manual for Courts-Martial
 (2) Mine Countermeasures
MCOTEA
 Marine Corps Operational Test and Evaluation Activity
MCP
 (1) Military Construction Plan
 (2) Military Construction Program
 (3) Mission Coordinating Paper
MCPDM
 Marine Corps Program Decision Meeting
MCPON
 Master Chief Petty Officer of the Navy
MCRBBS
 Marine Corps Reserve Bulletin Board System

MCRDAC
   Marine Corps Research, Development and Acquisition Command
MCRSC
   Marine Corps Reserve Support Center
MCS
   Maneuver Control System (USA)
MCSF
   (1) Marine Corps Security Force
   (2) Mobile Cryptologic Support Facility
MCSSD
   Mobile Combat Service Support Detachment (USMC)
MCSYSCOM
   Marine Corps System Command
MCTL
   Military Critical Technology List
MD
   McDonnell Douglas [firm]
MDA
   (1) Milestone Decision Authority
   (2) Missile Defense Act [1991]
MDAO
   Mutual Defense Assistance Office
MDAP
   Mutual Defense Assistance Program
M-day
   Mobilization day
MDCS
   Maintenance Data Collection System (USN)
MDCT
   Mine Detection Laser Technology
MDL
   Military Demarcation Line
MDQS
   Management Data Query System
MDR
   Medium Data Rate
MDS
   (1) Mission Design Series [aircraft and missile designations] (DOD)
   (2) Mission Display System (USN)
   (3) Modular Decontaminating System (USA)

MDSS
MAGTF Decision-Support System
MDT
Mean Down Time
MDW
Military District of Washington (USA)
MDZ
Maritime Defense Zone (USN-USCG)
ME
(1) Manufacturing Engineering
(2) Middle East
MEA
Munitions Effectiveness Assessment
MEB
Marine Expeditionary Brigade [formerly MAB] (USMC)
MEC
Main Evaluation Center (USN)
MECH
Mechanized
MED
(1) Manipulative Electronic Deception
(2) Medical
(3) Mediterranean
MEDCAP
Medical Civic Action Project (USMC)
MEDCOM
Medical Command (USA)
MEDDAC
Medical Activities (USA)
MEDEVAC
Medical Evacuation
MEDSOM
Medical Supply Optical and Maintenance (USA)
MEECN
Minimum Essential Emergency Communications Network (USAF)
MEF
(1) Major Equipment File
(2) Marine Expeditionary Force [formerly MAF]
(3) Middle East Force
MELIOS
Mini Eyesafe Laser Infrared Observation Set (USA)

MEPCOM
U.S. Military Enlistment Processing Command
MEPES
Medical Planning and Execution System
MER
(1) Manpower Estimate Report
(2) Multiple Ejector Rack (USN)
MET
Medium Equipment Transporter
METL
Mission-Essential Task List (USA)
METSAT
Meteorological Satellite
METT-T
Mission, Enemy, Troops, Terrain, Time (USA)
MEU
Marine Expeditionary Unit [formerly MAU]
MEU-SOC
Marine Expeditionary Units—Special Operations Capable
MEWSG
Multisevice Electronic Warfare Support Group [originally Maritime Electronic Warfare Support Group] (NATO)
MEWSS
Mobile Electronic Warfare Support System (USMC)
MEZ
Missile Engagement Zone
MF2K
Medical Force 2000 (USA)
MFCS
Microprocessor Flight Control System
MFD
Multi-Function Display
MFDS
Modular Fuel Delivery Station [shipboard installation] (USN)
MFHBF
Mean Flight Hours Between Failure
MFL
(1) Major Force List
(2) Master Force List
MFO
Multinational Forces and Observers

MFP
   (1) Major Force Program
   (2) Material Fielding Plan
MFPF
   Minefield Planning Folder (USN)
MG
   (1) Machine Gun
   (2) Military Government
MHA
   Military Health Affairs
MHD
   Magnetohydrodynamic
MHE
   Material-Handling Equipment
MHIP
   Missile Homing Improvement Program
MHS
   Message-Handling System
Mi
   Mil' [Soviet-Russian aircraft designation]
MI
   Military Intelligence
MIA
   Missing In Action
MIB
   Military Intelligence Board
MICAD
   Multipurpose Integrated Chemical Agent Alarm (USA)
MICAP
   Mission Critical Parts (USAF)
MICLIC
   Mine Clearing Line Charge
MICOM
   Missile Command (USA)
MIDEASTFOR
   Middle East Force
MIDS
   Multifunctional Information Distribution System (see JTIDS)
MIF
   Maritime Interception Force

MIG
> (1) Mikoyan-Gurevich [Soviet-Russian aircraft designation; changed from *MiG* to all capitals by the design bureau in 1991]
> (2) slang for any Soviet-Russian fighter aircraft [*mig*]

MIGCAP
> Combat Air Patrol [against Soviet-Russian fighters; may be flown by other nations]

MILCON
> Military Construction [funding]

MILES
> Multiple Integrated Laser Engagement System [training system]

MILGP
> Military Group

MILPERS
> Military Personnel

MILSATCOM
> Military Satellite Communications

MILSCAP
> Military Standard Contract Administration Procedure

MILSPEC
> Military Specification

MILSTAMP
> Military Standard Transportation And Movement Procedures

MILSTAR
> Military Strategic and Tactical Relay System

MILSTD
> Military Standard

MILSTEP
> Military Supply and Transportation Evaluation Procedures

MILSTRAP
> Military Standard Transaction Reporting and Accounting Procedures

MILSTRIP
> Military Standard Requisitioning and Issue Procedures

MIMIC
> Microwave Tube and Microwave Monolithic Integrated Circuits

MINEX
> Mine Warfare Exercise (USN)

MIO
> (1) Marine Inspection Office (USCG)
> (2) Maritime Interception Operations (USCG)

MIOAC
 Military Intelligence Officer Advanced Course
MIP
 (1) Management Improvement Plan
 (2) Model Installation Program
MIPE
 Mobile Intelligence Processing Element
MIPR
 Military Interdepartmental Purchase Request
MIR
 Mishap Investigation Report (USN)
MIRACL
 Mid-Infrared Advanced Chemical Laser (USN)
MIRV
 Multiple Independently Targetable Reentry Vehicle
MIS
 Management Information System
MISTE
 Military Intelligence Special Training Element
MITT
 (1) Mobile Imagery Transmission Terminal
 (2) Mobile Integrated Tactical Terminal
MIUW
 Mobile Inshore Undersea Warfare (USN)
MIW
 Mine Warfare (USN)
Mk
 Mark [equipment designation]
MLA
 Military Liaison Assistant
MLC
 (1) Military Liaison Committee
 (2) Military Load Classification
MLDT
 Mean Logistics Delay Time
MLF
 Multilateral Force
MLG
 Main Landing Gear
MLO
 Military Liaison Office

MLR
   Medium-Lift Requirement [helicopter/VSTOL] (USMC)
MLRS
   Multiple Launch Rocket System
MLS
   (1) Microwave Landing System (USAF)
   (2) Multilevel Security
MLV
   Medium Launch Vehicle (USAF)
mm
   millimeter
MM
   Minuteman [missile]
MMA
   Major Maintenance Availability (USN-USCG)
MMC
   Material Management Center (USA)
MMEI
   Military Medicine Education Institute (DOD)
MMH/FH
   Maintenance Manhours per Flight Hours
MMH/OH
   Maintenance Manhours to Operating Hours
MMHS
   Mechanized Material Handling System (USAF)
MMP
   Master Mobilization Plan
MMS
   Mast Mounted Sight (USA)
MMSIP
   Maintenance Management Systems Improvement Project (USAF)
MMT
   Manufacturing Methods Technology
MMWG
   Military Mobilization Working Group
MNC
   Major NATO Commander
MND
   Mission Need Determination
MNF
   Multinational Force

MNFP
  Multinational Fighter Program (NATO)
MNS
  (1) Mine Neutralization Systems (USN)
  (2) Mission Need Statement
MNV
  Mine Neutralization Vehicle (USN)
MO
  Medical Officer [British]
MOA
  Memorandum Of Agreement
MOB
  Main Operating Base
MOC
  Mobile Operations Center (USAF)
MOCC
  Mobile Operations Command Center
Mod
  Modification
MOD
  Ministry of Defense [British "Defence"]
MODE
  transportation mode
MODEM
  Modulator/Demodulator
MODLOC
  Miscellaneous Operational Details, Local Operations (USMC)
MOE
  Measure Of Effectiveness
MOI
  Marine Officer Instructor
MOMAG
  Mobile Mine Assembly Group (USN)
MOP
  (1) Magnetic Orange Pipe [minesweeping device] (USN)
  (2) Memorandum Of Policy (JCS)
MOPP
  Mission-Oriented Protection Posture
Mort
  Mortar

MOS
- (1) Maintenance Operations Section (USMC)
- (2) Military Occupational Specialty (USA)

MOSA
Minimum Operational Safe Altitude

MOSS
Mobile Submarine Simulator [decoy]

MOT
Military Ocean Terminal

MOU
Memorandum Of Understanding

MOUT
Military Operations in Urbanized Terrain (USA-USMC)

MOVREP
Movement Report (USN)

MP
- (1) Material Professional (USN)
- (2) Military Police

MPA
- (1) Main Political Administration of the Army and Navy [Soviet]
- (2) Main Propulsion Assistant (USN)
- (3) Maritime Patrol Aircraft [VP for patrol aircraft is more generally used in the United States]
- (4) Mission Payload Assessment (USAF)

MPC
Military Payment Certificates

MPCC
Minimum Professional Core Competence (USN)

MPES
Medical Planning and Execution System

MPF
- (1) Maritime Prepositioning Force
- (2) Multipurpose Facility

MPIM
Multi-Purpose Individual Munition (USA)

MPLH
Multipurpose Light Helicopter

MPM
Medical Planning Module

MPN
- (1) Manpower Personnel, Navy
- (2) Military Personnel Navy

M-POTS
　Mobile Psychological Operations Transmitter
MPS
　(1) Maritime Prepositioning Ship
　(2) Maritime Prepositioning Squadron
MPSM
　Multipurpose Submunitions
MPSRON
　Maritime Prepositioning Ship Squadron
MPT
　(1) Manpower and Training
　(2) Manpower, Personnel, and Training
MRAAM
　Medium-Range Air-to-Air Missile
MRASM
　Medium-Range Air-to-Surface Missile [Tomahawk variant; cancelled] (USAF)
MRB
　Material Review Board
MRBM
　Medium-Range Ballistic Missile
MRC
　Major Regional Contingency
MRCI
　Mine Readiness Certification Inspection (USN)
MRD
　Motorized Rifle Division [Soviet-Russian]
MRE
　Meals Ready-to-Eat
MRF
　(1) Missile Reconstitution Force (USAF)
　(2) Multi-Role Fighter [replacement for the F-16] (USAF)
MRG
　Movement Requirements Generator
MRL
　Multiple Rocket Launcher
MRP
　Material Requirements Planning
MRS
　Mobility Requirement Study
MRSA
　Material Readiness Support Agency

MRSI
Medium-Range SOF Insertion
MRSR
Multi-Role Survivable Radar (USA)
MRT
Miniature Receive Terminal
MRV
Multiple Reentry Vehicle
MS
(1) Medical Service Corps
(2) Milestone
M/S
Milestone
MSA
Morale Support Activity
MSAT-A
Multi-Sensor Aided Targeting—Air (USA)
MSB
Main Support Battalion (USA)
MSC
(1) Major Subordinate Command (NATO)
(2) Medical Service Corps (USA-USN)
(3) Military Sealift Command (USN)
MSD
Material Support Date
MSE
(1) Major Support Element
(2) Mobile Subscriber Equipment (USA)
MSFSG
Manned Space Flight Support Group (USAF)
MSG
Marine Security Guard
MSI
Multi-Spectral Imagery
MSIP
Multistage Improvement Program
MSL
(1) Mean Sea Level
(2) Missile
MSLEX
Missile Exercise

MSO
- (1) Marine Safety Office
- (2) Military Service Obligation

MSOW
Modular Standoff Weapon (NATO)

MSR
- (1) Magnetic Silencing Range (USN)
- (2) Main Supply Route

MSS
- (1) Mine Search System (USN)
- (2) Moored Surveillance System (USN)

MSSA
Master Safeguards and Security Agreements

MSS AS
Multistatic Sonar System Acoustic Source

MSSG
Marine Expeditionary Unit Service Support Group

MSTS
Military Sea Transportation Service [changed to MSC in 1970]

MT
Manufacturing Technology

MTACCS
Marine Tactical Command and Control System

MTBF
Mean Time Between Failure

MTBMA
Mean Time Between Maintenance Actions

MTBO
Mean Time Before Obsolescence (USN)

MTCR
Missile Technology Control Regime [1987]

MTF
- (1) Medical Treatment Facility
- (2) Message Text Format

MTI
Moving Target Indicator

MTL
Materials Technology Laboratory (USA)

MTMC
Military Traffic Management Command (USA)

MTM/D
Million-Ton-Miles per Day

MTO
  Mission Type Order
MTOE
  Modified Tables of Organization and Equipment
MTON
  Measurement Ton
MTR
  Military Technical Revolution
MTT
  Mobile Training Team
MTTR
  Mean Time To Repair
MTV
  Medium Tactical Vehicle
MULE
  Modular Universal Laser Equipment [training system]
MUSARC
  Major U.S. Army Reserve Command
MUSLO
  Morocco-U.S. Liaison Office
MUST
  Medical Unit, Self-contained, Transportable
MV
  Merchant Vessel [commercial merchant ship]
MVD
  Ministry of Internal Affairs [Soviet]
MWD
  Military Working Dog
MWF
  Medical Working File
MWR
  (1) Millimeter-Wave Radar
  (2) Moral, Welfare, and Recreation
Mya
  Myascheiv [Soviet-Russian aircraft designation]
MYP
  Multiyear Procurement

N

N1
> Staff Officer for Administration (USN)

N2
> Staff Officer for Intelligence (USN)

N3
> Staff Officer for Operations (USN)

N4
> Staff Officer for Supply/Logistics (USN)

N5
> Staff Officer for Plans (USN)

N6
> Staff Officer for Communications (USN)

N-( )*
> Principal directorates of the staff of the Chief of Naval Operations [from 1992; replacing the Office of the Chief of Naval Operations]

NAAS
> Naval Auxiliary Air Station

NAASW
> Non-Acoustic Anti-Submarine Warfare

NAAWS
> NATO Anti-Air Warfare System

NAB
> Naval Amphibious Base

NAC
> (1) Naval Avionics Center
> (2) North Atlantic Council (NATO)

NACA
> National Advisory Committee on Aeronautics [obsolete, became NASA]

NACC
> North Atlantic Cooperation Council

NACES
> Navy Aircrew Escape System

NADC
> Naval Air Development Center

NADEP
> Naval Aircraft Depot

NADOC
　Naval Aviation Depot Operations Center
NAE
　Navy Acquisition Executive
NAEC
　Naval Air Engineering Center
NAEW
　NATO Airborne Early Warning [aircraft]
NAF
　(1) Naval Air Facility
　(2) Nonappropriated Fund (USN)
　(3) Numbered Air Force (USAF)
NAI
　Named Areas of Interest
NALO
　Naval Air Logistics Office
NAOTS
　Naval Aviation Ordnance Test Station
NAP
　(1) Naval Aviation Pilot (USN)
　(2) Naval Aviation Plan (USN)
NAPDD
　Non-Acquisition Program Definition Document (USN)
NAPR
　NATO Armaments Planning Review
NAPS
　Naval Academy Preparatory School
NAR
　Naval Air Reserve
NARDIC
　Navy Acquisition, Research, and Development Information Center
NARF
　Naval Air Rework Facility (USN)
NAS
　(1) National Academy of Sciences
　(2) National Airspace System (USAF)
　(3) Naval Air Station
　(4) Naval Audit Service
NASA
　National Aeronautics and Space Administration

NASP
National Aerospace Plane [X-30]
NASSCO
National Steel and Shipbuilding Company [San Diego, Calif.]
NATC
Naval Air Test Center [Patuxent River, Md.]
NATF
Navy Advanced Tactical Fighter
NATO
North Atlantic Treaty Organization
NATOPS
Naval Air Training and Operating Procedures Standardization [program]
NATS
Naval Air Transport Service
NAUS
National Association of Uniformed Services
NAUTIS-F
Naval Autonomous Information System—Frigate
NAVAIR
Naval Air Systems Command
NAVAIRLANT
Naval Air Force Atlantic
NAVAIRPAC
Naval Air Force Pacific
NAVAIR SYSCOM
Naval Air System Command
NAVCAD
Naval Aviation Cadet
NAVCENT
Naval Forces, U.S. Central Command [see USNAVCENT]
NAVCOMMSTA
Naval Communications Station
NAVEUR
Naval Forces, Europe
NAVFAC
(1) Naval Facilities Engineering Command
(2) Naval Facility
NAVFAC SYSCOM
Naval Facilities Engineering Command

NAVFOR
 Naval Forces
NAVINTCOM
 Naval Intelligence Command
NAVIXS
 Navy Information Transfer System
NAVMAT
 Naval Material Command [abolished]
NAVMIC
 Naval Maritime Intelligence Center [formerly NISC and then NTIC]
NAVNET
 Navy Network
NAVOCFORMED
 Naval On-Call Force, Mediterranean
NAVRESFOR
 Naval Reserve Force
NAVRESO
 Navy Resale Systems Office
NAVRESOFSO
 Navy Resale System Field Support Office
NAVRESSOFO
 Navy Resale and Service Support Office, Field Support Office
NAVSAFECEN
 Naval Safety Center [Norfolk, Va.]
NAVSAT
 Navigation Satellite
NAVSEA
 Naval Sea Systems Command
NAVSEA SYSCOM
 Naval Sea Systems Command
NAVSEC
 Naval Ship Engineering Center [merged with Naval Sea Systems Command in 1979]
NAVSECGRU
 Naval Security Group
NAVSPACCOM
 Naval Space Command
NAVSPECWAR
 Naval Special Warfare
NAVSPECWARCOM
 Naval Special Warfare Command

NAVSPECWARGRU
Navy Special Warfare Group
NAVSTA
Naval Station
NAVSTAR
Navigation Satellite Timing and Ranging [satellite]
NAVSTAR GPS
Navigation Satellite Timing and Ranging Global Positioning System
NAVSTKWARCEN
Naval Strike Warfare Center
NAVSUP
Naval Supply Systems Command
NAVSUP SYSCOM
Naval Supply System Command
NAWC
Naval Air Warfare Center
NBC
Nuclear, Biological, and Chemical
NBCC
Nuclear, Biological, and Chemical Contamination
NBCRS
Nuclear, Biological, and Chemical Reconnaissance System (USA)
NBDL
Naval Biodynamics Laboratory
NBG
Naval Beach Group
NC
Nurse Corps (USN)
NCA
National Command Authority
NCAA
NATO Civil Air Augmentation
NCC
Naval Command College
NCCO
Navy Command and Control System
NCCOSC
Naval Command, Control, and Ocean Surveillance Center [formerly NOSC and other activities]
NCCS
Navy Command and Control System

NCMC
NORAD Cheyenne Mountain Complex
NCMP
Navy Capabilities and Mobilization Plan
NCO
(1) Net Control Outstation
(2) Non-Commissioned Officer
NCOES
Noncommissioned Officer Education System (USA)
NCOIC
Non-Commissioned Officer-In-Charge (USAF)
NCOLP
Non-Commissioned Officer Logistics Program
NCP
Non-U.S. Coalition Partner
NCPS
Nuclear Contingency Planning System
NCR
National Capital Region
NCR SPTG
National Capital Regional Support Group (USAF)
NCS
(1) National Communication System
(2) Naval Control of Shipping
(3) Net Control Station (USA)
NCSC
Naval Coastal Systems Center [Panama City, Fla.]
NCSO
Naval Control of Shipping Organizations
NCTAMS
Naval Computer and Telecommunications Area Master Station
NCTC
Naval Computer and Telecommunications Command
NCTR
Non-Cooperative Target Resolution
NCUA
National Credit Union Administration
ND
Normal Deployment posture
NDB
Nuclear Depth Bomb
NDDS
Nuclear Detonation Detection System

NDI
Nondevelopment Item
NDMS
National Disaster Medical System
NDP
Naval Doctrine Publication
NDRF
National Defense Reserve Fleet
NDRI
National Defense Research Institute
NDS
(1) National Defense Stockpile
(2) Nuclear Detonation Detection System
(3) NUDET Detection System (USAF)
ND/SB
Nuclear Depth/Strike Bomb
NDSF
National Defense Sealift Fund
NDSM
National Defense Service Medal
NDTA
National Defense Transportation Association
NDU
National Defense University [Washington, D.C.]
NEACP
National Emergency Airborne Command Post
NEARTIP
Near-Term Improvement Program [Mk 46 torpedo]
NEC
(1) Naval Enlistment Classification
(2) Navy Enlisted Codes
NEO
Non-combat Evacuation Operation
NEPA
National Environmental Policy Act (EPA)
NESP
Navy EHF Satellite Program
NET
New Equipment Training (USA)
NETT
New Equipment Training Team (USA)

NFAF
　Naval Fleet Auxiliary Force
NFC
　Numbered Fleet Commander
NFIP
　National Foreign Intelligence Program
NFO
　Naval Flight Officer
NFOIO
　Navy Field Operational Intelligence Office [changed to NOIC in 1984]
NFR
　NATO Frigate [NFR-90]
NG
　National Guard [ANG or ARNG]
NGB
　National Guard Bureau
NGF
　Naval Gunfire
NGFS
　Naval Gunfire Support
NH
　NATO Helicopter [NH-90]
NIA
　Naval Intelligence Activity
NIC
　(1) National Intelligence Council
　(2) Naval Intelligence Command
NICI
　National Interagency Counterdrug Institute [Camp San Luis Obispo, Calif.]
NICP
　National Inventory Control Point (USA)
NICS
　NATO Integrated Communications System
NID
　Naval Intelligence Database
NIE
　National Intelligence Estimate
NIF
　Navy Industrial Fund

NIPS
- (1) Naval Intelligence Processing System
- (2) NMCS Information Processing System

NIS
- (1) NATO Identification System [IFF system]
- (2) Naval Investigative Service

NISC
Naval Intelligence Support Center [later NTIC; now NAVMIC]

NISP
NUWEP Intelligence Support Plan

NITF
National Imagery Transmission Format

NJP
Nonjudicial Punishment

NLG
Nose Landing Gear

NLMS
Navigational Lane Marking System (USN)

NLOS
Non–Line-Of-Sight [anti-tank missile] (USA)

NLOS-R
Non–Line-Of-Sight—Rear (USA)

NLS
National Launch System

NLSF
Navy Logistics Support Force

NMC
- (1) Naval Material Command [abolished]
- (2) Network Management Centers
- (3) Not Mission Capable (USN)

NMCB
Naval Mobile Construction Battalion

NMCC
National Military Command Center

NMCS
National Military Command System

NME
Naval Material Establishment

NMIA
National Military Intelligence Association

NMIC
(1) National Maritime Intelligence Center [created in 1992 from the Naval Intelligence Center and other intelligence activities in the Washington, D.C., area] (USN)
(2) National Military Intelligence Center [DIA]
NMIST
National Military Intelligence Support Team [DIA]
NMITC
Navy and Marine Corps Intelligence Training Center
NMPC
Naval Military Personnel Command
NMS
National Military Strategy
NMSA
Nonnuclear Munitions Storage Area (USAF)
NMSD
National Military Strategy Document
NMW
Naval Mine Warfare
NNBIS
National Narcotics Border Interdiction System
NNMC
National Naval Medical Center [Bethesda, Md.]
NNOR
Non-Nuclear Ordnance Requirement (USN)
NNS
(1) Navy News Service
(2) Newport News Shipbuilding
NOAA
National Oceanic and Atmospheric Administration
NOB
Naval Operating Base
NOBC
Naval Officer Billet Classification
NODIS
No Distribution [security restriction]
NOE
Nap-Of-Earth [helicopter flight profile]
NOFORN
Not Releasable to Foreign Nationals
NOGS
Night Observation Gunship System

NOIC
  (1) Naval Ocean Intelligence Center
  (2) Navy Operational Intelligence Center [now Naval Maritime Intelligence Center (NAVMIC)]
NOLO
  No Live Operator (USN)
NOP
  Nuclear Operations
NOPC
  Naval Oceanographic Processing Center
NOPF
  National Oceanographic Processing Facility
NOPLAN
  No Plan [Available or Prepared]
NORAD
  North American Aerospace Defense Command [joint U.S.-Canadian command]
NORCANUKUS
  Norway, Canada, United Kingdom, United States
NORDO
  No Radio [aircraft] (USN)
NOREX
  Nuclear Operational Readiness Exercise (USN)
NORS
  Nonoperational Ready (USN)
NORTHAG
  Northern Army Group, Central Europe (NATO)
NORTIC
  NORAD Technical Intelligence Center
NOS
  Night Observation System
NOSC
  Naval Ocean Systems Center [now component of NCCOSC]
NOSIC
  Naval Ocean Surveillance Information System (USN)
NOTAM
  (1) Notice to Airmen (USN)
  (2) Notice to Mariners
NOTAR
  No Tail Rotor

NOTS
>(1) Naval Ocean Transport Service [changed to MSTS in 1949 now MSC]
>(2) Naval [aviation] Ordnance Test Station

NPB
>Neutral Particle Beam

NPDM
>Navy Program Decision Meeting

NPE(S)
>Nuclear Planning and Execution (Service)

NPFL
>National Patriotic Front of Liberia

NPG
>(1) Nonunit Personnel Generator
>(2) Nuclear Planning Group (NATO)

NPGS
>Naval Postgraduate School [Monterey, Calif.]

NPIC
>National Photographic Interpretation Center

NPL
>National Priorities List [Environmental Protection Agency]

NPR
>New Production Reactors [program]

NPS
>Nonprior Service

NPT
>Non-Proliferation Treaty [1968]

NPWIC
>National Prisoner of War Information Center

NRAC
>Naval Research Advisory Committee

NRC
>(1) Non–unit-Related Cargo
>(2) Nuclear Regulatory Committee

NRDC
>Natural Resources Defense Council (private)

NRF
>Naval Reserve Force

NRL
>Naval Research Laboratory [Washington, D.C.]

NRO
>National Reconnaissance Office [changed to CIO in 1992]

NROSS
  Navy Remote Ocean Sensing System [satellite]
NROTC
  Naval Reserve Officers Training Corps
NRP
  Non–unit-Related Personnel
NRRC
  Nuclear Risk Reduction Center
NRSO
  Navy Resale Systems Office
NRT
  (1) Naval Reserve Training [now NRF]
  (2) Near-Real Time
NS
  Naval Station
NSA
  National Security Agency [Fort George G. Meade, Md.]
NSB
  Naval Studies Board [National Academy of Sciences]
NSC
  (1) National Security Council
  (2) Naval Staff College
NSCCA
  Nuclear Safety Cross-Check Analysis
NSCID
  National Security Council Intelligence Directive
NSDAB
  Non–Self-Deployable Aircraft and Boats (USN)
NSDD
  National Security Decision Directive
NSE
  Naval Support Element
NSEP
  National Security Emergency Preparedness
NSF
  (1) National Science Foundation
  (2) Navy Stock Fund
NSL
  Naval Submarine League
NSN
  National Stock Number

NSNF
　Non-Strategic Nuclear Forces
NSO
　Non-SIOP Options
NSOC
　(1) National Signals Intelligence Operations Center (NSA)
　(2) Navy Satellite Operations Center
NSP
　Navy Support Plan
NSPG
　National Security Planning Group
NSRDC
　Naval Ship Research and Development Center [formerly David W. Taylor NSRDC; Carderock, Md.]
NSS
　National Supply System
NSSDC
　National Space Science Data Center (NASA)
NSSG
　NATO/SHAPE Support Group
NSSM
　NATO Sea Sparrow Missile
NSSN
　New Nuclear Attack Submarine
NST
　(1) Navy Standard Teleprinter
　(2) Nuclear and Space Talks
NSTL
　National Strategic Target List
NSTT
　Naval Strategy Think Tank
NSW
　Naval Special Warfare
NSWC
　Naval Surface Weapons Center
NSWO
　Nuclear Surface Warfare Officer (USN)
NSWP
　Non-Soviet Warsaw Pact
NSWTG
　Naval Special Warfare Task Group

NSWTG-CENT
Naval Special Warfare Task Group, Central Command
NSWU
Naval Special Warfare Unit
NSY
Naval Shipyard [also NSYd]
NTC
National Training Center [Fort Irwin, Calif.] (USA)
NTCC
Naval Telecommunications Center
NTCS-A
Navy Tactical Command System Afloat
NTCS-A (DM)
Navy Tactical Command System Afloat—Database Management
NTCS-A (SO)
Navy Tactical Command System Afloat—Staff Officer Course
NTDS
Naval Tactical Data System
NTIC
Naval Technical Intelligence Center [pronounced *N-tech*; formerly NISC, now NAVMIC]
NTIS
National Technical Information Service [Dept. of Commerce]
NTM
National Technical Means [arms verification]
NTP
(1) Naval Training Publication
(2) Navy Training Plan
NTPF
Near-Term Prepositioning Force
NTPI
Nuclear Training Proficiency Inspection (USN)
NTPOC
Navy Technical Point Of Contact
NTPS
Near-Term Prepositioning Ship
NTU
New Threat Upgrade (USN)
NUCWEPS
Nuclear Weapons Employment Course

NUDET
　　Nuclear Detonation
nuke**
　　(1) nuclear or nuclear-trained personnel [slang]
　　(2) nuclear weapon [slang]
NUPOC
　　Nuclear Power Candidate (USN)
NUSC
　　Naval Underwater Systems Center [New London, Conn.]
NUSN
　　Non-U.S. NATO
NVG
　　Night Vision Goggles
NVGS
　　Night Vision Goggle Sensor
NVR
　　Naval Vessel Register
NVS
　　Night Vision System (USA)
NWC
　　(1) National War College [Washington, D.C.]
　　(2) Naval War College [Newport, R.I.]
　　(3) Naval Weapons Center
　　(4) Nuclear Weapons Center
NWF
　　Naval War College Foundation
NWL
　　Naval Weapons Laboratory
NWOC
　　Naval Weather and Oceanographic Center
NWP
　　Naval Warfare Publication
NWS
　　National Weather Service
NWSC
　　Naval Weapons Support Center
NWTDB
　　Naval Warfare Tactical Data Base

oa
>overall [ship length]

OA
>Obligation Authority

OAB
>Outer Air Battle (USN)

OADR
>Originating Agency's Determination Required [declassification]

OAS
>(1) Offensive Attack System
>(2) Offensive Avionics System
>(3) Office of the Assistant Secretary
>(4) Organization of American States

OASD SO/LIC
>Office of the Assistant Secretary of Defense for Special Operations and Low Intensity Conflict

OASIS
>Over-the-Horizon Airborne Sensor Information System (USN)

OAST
>Overland Air Superiority Training (USN)

OAT
>Outside Air Temperature

OB
>(1) Operating Budget
>(2) Order of Battle [preferred; also OOB]

OBA
>Oxygen Breathing Apparatus (USN)

OBE
>Overtaken-By-Events [slang]

OBEWS
>On-Board Electronic Warfare System

OBIGGS
>On-Board Inert Gas Generating System (USAF)

OBU
>Ocean Surveillance Informations System Baseline Upgrade

O/C
>Observer/Controller

OCA
 Offensive Counter Air
OCAC
 Operations, Control, and Analysis Center
OCCA
 Ocean Cargo Clearance Authority
OCHAMPUS
 Office CHAMPUS
OCIT
 Office of Combat Identification Technology
OCLL
 Office, Chief of Legislative Liaison (USA)
OCO
 Office, Chief of Ordnance (USA)
OCONUS
 (1) Outside the Continental United States
 (2) Overseas CONUS [Hawaii, Alaska]
OCS
 Officer Candidate School
OCSA
 Office of the Chief of Staff, U.S. Army
OCT
 Operational Climatic Testing
ODA/ODIF
 Office Document Architecture/Office Document Interchange Format
ODC
 Office of Defense Cooperation
ODP
 Officer Distribution Plan (USN)
ODRP
 Office of Defense Representative—Pakistan
ODS
 Operating Differential Subsidies
ODT
 Overseas Deployment Training (USA)
OEA
 Office of Economic Adjustment
OEC
 Operational Evaluation Command (USA)
OECD
 Organization for Economic Cooperation and Development

OED
OSIS Evolutionary Development
OEG
Operational Evaluation Group (USN)
OER
Officer Evaluation Report (USA)
OFP
Operational Flight Program
OFPP
Office of Federal Procurement Policy [OMB]
OFT
Operational Flight Trainer (USN)
OGC
Office of General Counsel
OGE
Out-of-Ground-Effect [hover condition]
OGS
Overseas Ground Station
OH
Observation Helicopter (USA)
OHA
(1) Off-station Housing Allowance
(2) Overseas Housing Allowance
OHO
Ordnance Handling Officer (USN)
OI
Operating Income
OIBA
Office of Industrial Base Assessment
OIC
Officer In Charge
OICC
Operational Intelligence Crisis Center [DIA]
OIM
Office of Industrial Mobilization
OIP
Optical Improvement Program
OJCS
Organization of the Joint Chiefs of Staff
OJT
On-the-Job Training

OLA
Office of Legislative Affairs (USN)
OLF
Outlying Field (USN)
OLS
Optical Landing System (USN)
O&M
Operation and Maintenance
OMA
Operations and Maintenance, Army
OMB
(1) Office of Management and Budget
(2) Operational Maintenance Battalion (USA)
OMC
Office of Military Cooperation
OMFTS
Operational Maneuver From The Sea (USMC)
OMG
Operational Maneuver Group [Soviet-Russian]
O&MN
Operations and Maintenance, Navy
OMPF
Official Military Personnel File
ONC
On-Site Container
ONI
Office of Naval Intelligence
ONR
Office of Naval Research
ONT
Office of Naval Technology
OOA
Out Of Area [ship operations] (USN)
OOB
Order of Battle [OB preferred]
OOD
Officer Of the Deck (USN)
O&O PLAN
Operational and Organizational Plan (USA)
OOV
Objects Of Verification [arms control]

OP³
 Overt Peacetime Psychological Operations Program
OP
 (1) Observation Post
 (2) Other Procurement
OP-( )
 Former code for offices in the Office of the Chief of Naval Operations (OPNAV) [changed to N-series designations in 1992]
OPA
 Office of Program Appraisal [U.S. Secretary of the Navy]
OPCEN
 Operations Center
OPCOM
 Operational Command
OPCON
 Operational Control
OPDEC
 Operational Deception
OPDEC PL
 Operational Deception Planner
OPDEC PLN
 OPDEC Planning
OPDS
 Offshore POL Discharge System
OPEC
 Organization of Petroleum Exporting Countries
OPEVAL
 Operational Evaluation (USN)
OPINTEL
 Operational Intelligence
OPLAN
 Operational Plan
OPM
 Office of Personnel Management
OPMD
 Officer Personnel Management Directorate (USA)
OPMS
 Officer Personnel Management System (USA)
OPN
 Other Procurement, Navy [funding term]

OPNAV
Office of the Chief of Naval Operations [changed to Chief of Naval Operations staff in 1992]

OPNAVINST
OPNAV Instruction (USN)

OPORD
Operation Order

OPP
Offload Preparation Party (USMC)

OPPA
Operation Plan Package Appraisal

OPPE
Operational Propulsion Plant Examination (USN)

OPR
Office of Primary Responsibility

OPREP
Operational Report

Ops
Operations

OPSEC
Operations Security

OPSG
Operation Plans Steering Group

OPS O
Operations Officer (USN)

OPT
Operations Planning Team (USAF)

OPTAR
Operating Target (USN)

OPTASK
Operational Tasking

OPTEC
Operational Test and Evaluation Center (USA)

OPTEMPO
Operating Tempo (USN)

OPTEVFOR
Operational Test and Evaluation Force (USN)

OPV
Offshore Patrol Vessel

OQQ
Officer Qualification Questionnaire (USN)

OR
- (1) Operational Readiness
- (2) Operational Requirement (USN)
- (3) Operations Research

ORB
Officer Record Brief (USA)

ORD
Operational Requirements Document

ORDALT
Ordnance Alteration (USN)

ORE
Operational Readiness Evaluation (USN)

ORF
Operational Readiness Float (USN-USMC)

ORG
Origin

ORR
Operational Readiness Review

OR/SA
Operations Research/Systems Analysis

ORSE
Operational Reactor Safeguard Examination (USN)

ORTS
Operational Readiness Test System (USN)

ORWG
Operational Requirements Working Group

O/S
Operations and Support Phase

OS
Operational Suitability

O&S
Operations and Support

OSA
Operational Support Airlift (USAF)

OSC
Objective Supply Capability (USA)

OSD
Office of the Secretary of Defense

OSHA
Occupational Safety and Health Administration

OSI
Open Systems Interconnection

OSIA
On-Site Inspection Agency (DOD)
OS/IPC
Operating System/Inter-Process Communications
OSIS
(1) Ocean Surveillance Information Center (USN)
(2) Ocean Surveillance Information System (USN)
OSO
Officer Selection Officer (USMC)
OSP
(1) Offshore Procurement (USN)
(2) Ocean Surveillance Product
OSS
(1) Ocean Surveillance Satellite
(2) Ocean Surveillance System
(3) Operations Support System
OSTP
Office of Science and Technology (White House)
OT
Operational Test
OTA
(1) Office of Technology Assessment [Congress]
(2) Operational Test Agency
OTC
Officer in Tactical Command (USN)
OTCIXS
Officer-in-Tactical-Command Information Exchange System (USCG-USN)
OT&E
Operational Test and Evaluation
OT/FT
Operational Test/Follow-on Test [missiles]
OTH
Over-The-Horizon
OTH-B
Over-The-Horizon Backscatter [radar]
OTH-T
Over-The-Horizon Targeting
OTMS
Operational Technical Management System (USN)
OTP
Outline Test Plan

OTRA
Other-Than-Regular-Army
OTS
Officer Training School (USAF)
OTSG
Office of The Surgeon General
OTU
Operational Training Unit [aviation] (British)
OUSD(A)
Office of the Under Secretary of Defense (Acquisitions)
OUT
Outsized Cargo (USAF)
OUTS
Operational Unit Transportable System
OUTUS
Outside Continental U.S. (USN)
OVR
Oversized Cargo (USAF)

P

P³I
Pre-Planned Product Improvement
PA
(1) Precision-Acrobatics
(2) Product Assurance
(3) Program Authorization (USAF)
(4) Public Affairs
P&A
Price and Availability
PAA
Primary Aircraft Authorized (USAF)
PAC
(1) Pacific
(2) Personnel and Administrative Center

PACAF
Pacific Air Forces (USAF)
PACE
Program for Afloat College Education (USN)
PACOM
Pacific Command
PACS
Programmable Armament Control Set
PA&E
Program Analysis and Evaluation
PALS
Permissive Action Link System
Pam
Pamphlet
PAM
(1) Payload Assist Module (USAF)
(2) Procurement Aircraft and Missiles
PAO
Public Affairs Office[r]
PAPS
Periodic Armaments Planning System (NATO)
PAR
(1) Parameter Acquisition Radar
(2) Phased Array Radar
(3) Precision Approach Radar (USN)
(4) Program Assessment Review (USAF)
PARCS
Perimeter Acquisition Radar Attack Characterization System
PARPRO
Peacetime Airborne Reconnaissance Program (USAF)
PARR
Program Analysis and Resource Review (USA)
PAS
Primary Alerting System
PASSEX
Passing Exercise (USN)
PAT
(1) Patrol
(2) Process Action Team (USA)
PAT&E
Production Acceptance Test and Evaluation

PATS
Primary Aircraft Training System (USAF-USN)
PAVE PAWS
Precision Acquisition of Vehicle Entry and Phased Array Warning System (USAF)
PAX
Passengers (USAF)
PB
(1) President's Budget
(2) Program Baseline
PBD
Program Budget Decision
PBV
Post Boost Vehicle
PCA
Physical Configuration Audit
PCAD
Program Change Approval Document (USN)
PCCADS
Panoramic Cockpit Control And Display System (USAF)
PCD
Program Change Decision
PCE
Professional Continuing Education (USAF)
PCL
Pocket Check List [aviation] (USN)
PC-LITE
Processor, Laptop Imagery Transmission Equipment
PCM
Program Cost Management
PCO
(1) Procuring Contracting Officer
(2) Prospective Commanding Officer (USN)
PCR
(1) Procurement Center Representative
(2) Program Change Request
PCS
Permanent Change of Station
PD
(1) Position Description
(2) Probability of Detection
(3) Program Director (USAF)

PDA
 (1) Principal Decision Authority
 (2) Principal Developing Agency
PDD
 Package Designation and Description File
PDES
 Product Data Exchange Specification
PDM
 (1) Program Decision Memorandum
 (2) Programmed Depot Maintenance
PDMS
 Point Defense Missile System (USN)
PDP
 Program Development Plan
PDR
 Preliminary Design Review
PDRC
 Program Development Review Committee (USN)
PDS
 Personnel Data System
PDSS
 Post Deployment Software Support
PDU
 Pilots Display Unit (USA)
PE
 (1) Planning Estimate
 (2) Procurement Executive
 (3) Program Element [budget term]
PEB
 Propulsion Examination Board (USN)
PEBD
 Pay Entry Base Date (USN)
PEC
 (1) Program Element Code
 (2) Program Evaluation Group (USMC)
PECC
 Pacific Economic Cooperation Conference
PECI
 Productivity Enhancing Capital Investment
PEIS
 Programmatic Environmental Impact Statement

PEM
Program Element Monitor (USAF)
PENAID
Penetration Aid [missile]
PEO
Program Executive Officer (DOD)
PEP
(1) Producibility Engineering and Planning
(2) Producibility Enhancement Program
PERMS
Personnel Electronic Records Management System (USN-USAF)
PERS
Personnel
PERSCOM
Personnel Command (USA)
PERT
Program Evaluation Review Technique
PESO
Product Engineering Services Office
PFF
Planning Factors File
PFM
Program Financial Management
PFR
Pulse Repetition Frequency
PFT
Physical Fitness Test
PGC
Policy Guidance Council [DSMC]
PGIP
Post Graduate Intelligence Program
PGM
(1) Planning Guidance Memorandum
(2) Precision Guided Munitions
PGT
Platoon Gunnery Trainer (USA)
PHIB
Amphibious (USN-USMC)
PHIBEX
Amphibious Exercise
PHIBLEX
Amphibious Landing Exercise (USN-USMC)

PHS
 Public Health Service
PHST
 Packaging, Handling, Storage, and Transportation
PI
 Product Improvement
PIC
 (1) Parent Indicator Code
 (2) Pilot In Command (USN)
PID
 Plan Identification Number
PIE
 Pyrotechnically Initiated Explosive
PIF
 Productivity Investment Fund
PIM
 (1) Planned Incremental Modernization
 (2) Position of Intended Movement (USN)
PIMA
 Portable Intelligence Maintenance Aid (USA)
PIN
 Personnel Increment Number
PINS
 Precise Integrated Navigation System (USN)
PIO
 Public Information Officer
PIOB
 President's Intelligence Oversight Board
PIP
 (1) Productivity Improvement Program
 (2) Productivity Improvement Proposal
PIR
 Priority Intelligence Requirements (USA)
PIRAZ
 Positive Radar Identification Advisory Zone
pK
 Probability of Kill
PKG-POL
 Packaged POL
PKO
 Peacekeeping Operations

PL
    (1) Phase Line (USA)
    (2) Public Law
PLA
    Peoples Liberation Army [China]
Plat
    Platoon [also Plt]
PLAT
    Pilot Landing Aid Television (USN)
PLGR
    Precise Lightweight Global Positioning System Receiver
PLL
    Prescibed Load List
PLRS
    Position Location Reporting System
PLRS/JTIDS
    Position Location Reporting System/Joint Tactical Information Distribution System (USA)
PLS
    Palletized Load System
PLSS
    Precision Location Strike System
Plt
    Platoon [also Plat]
PM
    (1) Preventive Maintenance
    (2) Product Manager
    (3) Program Manager
    (4) Project Manager
    (5) Provost Marshal
PMA
    (1) Phased Maintenance Availability (USN)
    (2) Probability of Mission Abort
PMC
    (1) Partially Mission Capable (USA-USN)
    (2) Procurement, Marine Corps [funding]
PMCS
    (1) Professional Military Comptroller School (USN)
    (2) Program Management Control System
PMD
    (1) Program Management Directive (USAF)
    (2) Program Management Document

PMDB
　Program Management Decision Brief [DSMC]
PME
　(1) Primary Mission Equipment (USA)
　(2) Professional Military Education
PMJEG
　Performance Measurement Joint Executive Group
PMO
　(1) Program Management Office
　(2) Project Management Office (USA)
PMP
　Program Management Plan
PMR
　(1) Pacific Missile Range
　(2) Primary Mission Readiness [flight hours] (USN-USMC)
　(3) Program Management Review
PMRT
　Program Management Responsibility Transfer (USAF)
PMS
　(1) Pedestal Mounted Stinger [missile] (USA)
　(2) Planned Maintenance Subsystem (USN)
PMSS
　Program Manager's Support System
PMTC
　Pacific Missile Test Center [Point Mugu, Calif.]
PNVS
　(1) Pilot Night Vision Sensor
　(2) Pilot Night Vision System
POA&M
　Plan of Action and Milestones
POC
　(1) Point Of Contact
　(2) Privately Owned Conveyance (USN)
POD
　Port Of Debarkation
POE
　(1) Port Of Embarkation
　(2) Projected Operational Environment
POG
　Psychological Operations Group
POL
　Petroleum, Oils, and Lubricants

POLAD
 Political Adviser
POM
 (1) Preparation for Overseas Movement (USN)
 (2) Program Objective Memorandum
POMCUS
 Prepositioning Of Material Configured to Unit Sets
POP
 (1) Paperless Ordering Placement [system]
 (2) Proof Of Principle (USA)
PORSE
 Post-Overhaul Reactor Safeguards Examination (USN)
PORTS
 (1) Portable Remote Telecommunications System
 (2) Port Characteristics File
POS
 (1) Peacetime Operating Stocks
 (2) Port(s) Of Support
POSF
 Port Of Support File
POST
 Prototype Ocean Surveillance Terminal (USN)
POT&I
 Pre-Overhaul Testing and Inspection (USN)
POV
 Privately Owned Vehicle
POW
 Prisoner Of War
PPAC
 Product Performance Agreement Center (USAF)
PPBES
 Planning, Programming, Budgeting and Execution System (USA)
PPBS
 Planning, Programming, and Budgeting System (DOD)
PPL
 Provisioning Parts List
PPP
 (1) Prepositional Procurement Package
 (2) Priority Placement Program
PPS
 Post-Production Support

PQS
Personnel Qualification Standards
PR
Procurement Request
PRAM
(1) Primary Report of Aircraft Mishap (USA)
(2) Productivity, Reliability, Availability, and Maintainability (USAF)
PRAMPO
Productivity, Reliability, Availability, and Maintenance Program Office (USAF)
PRAT
Production Reliability Acceptance Test
PRC
(1) Peoples Republic of China
(2) Program Review Committee (USAF)
Prcht
Parachute (USA)
PRD
Projected Rotation Date (USN)
PRDA
Program Research and Development Announcement (USAF)
PREPRO
Prepositioning ship (USN)
PRESCOM
Personnel Command (USA)
PRG
(1) Peacekeeper Rail Garrison [cancelled 1991] (USAF)
(2) Program Review Group
PRIME
Precision Range Integrated Maneuver Exercise (USA)
PRIME BEEF
Priority Improvement Management Effort Base Engineering Emergency Force (USAF)
PRIME RIBS
Priority Improvement Management Effort Readiness In Base Services (USAF)
PRIMUS
Primary Medical Care for the Uniformed Services
PRM
Presidential Review Memorandum

PRMP
Plutonium Recovery Modification Project
PRO
Plant Representative Office
PROBASE
Production Base Information System [formerly DINET]
PROD
Production
PROD/DEPL
Production and Deployment Phase
PROM
Programmable Read-Only Memory
PROV
Provisional
PROVORG
Providing Organization
PRR
Production Readiness Review
PSA
(1) Personnel Support Activity
(2) Port Support Activity
(3) Post-Shakedown Availability (USN)
PS&A
Pharmacy, Supply, and Administration
PSC
Principal Subordinate Command (NATO)
PSD
(1) Personnel Support Detachment
(2) Propulsion System Demonstrator (USMC)
PSE
Peculiar Support Equipment
PSHD
Port Security Harbor Defense
PSI
Personnel Security Investigation
PSM
Professional Staff Member [Congress]
PSS
Personnel Service Support (USA)
PSSD
Personnel Service Support Directorate

PSU
 Port Security Unit (USCG)
PSV
 Pseudo-Synthetic Video
PSYOPS
 Psychological Operations
PT
 Physical Training
P&T
 Personnel and Training
PUFFS
 Passive Underwater Fire-Control Feasibility System
PUP
 Performance Update Program (USAF)
PVO
 Air Defense Forces [Voyska Protivovozdushnoy Oborony, Soviet-Russian]
PW
 Prisoner of War
P&W
 Pratt and Whitney [United Technologies]
PWB
 Printed Wiring Board
PWF
 Personnel Working File
PWG
 POM Working Group (USMC)
PWIC
 Prisoner of War Information Center
PWIS2
 Prisoner of War Information System
PWP
 White Phosphorus [ammunition]
PWR
 Pressurized Water Reactor
PWRM
 Pre-positioned War Reserve Material
PWRMS
 Pre-positioned War Reserve Material Stocks
PWRS
 Pre-positioned War Reserve Stocks

PX
  Post Exchange
PXO
  Prospective Executive Officer (USN)
PY
  Prior Year [fiscal]
PZ
  Pickup Zone (USA)

Q

QA
  Quality Assurance
QAR
  Quality Assurance Representative
QBL
  Qualified Bidders List
QC
  Quality Control
QCI
  Quality Control Inspection
QCR
  Qualitative Construction Requirement
QE2
  AN/BQQ-5E [sonar installation]
QM
  Quartermaster (USA)
QMB
  Quality Management Board
QMC
  Quartermaster Corps (USA)
QMP
  Qualitative Management Program
QPL
  Qualified Products List

QQPRI
   Qualitative and Quantitative Personnel Requirements Information (USA)
QRC
   Quick Reaction Capability
QRCC
   Quick Reaction Combat Capability
QRF
   Quick Reaction Force
QT
   Qualification Test
QUICKTRANS
   Contract Airlift Service (USA)
Q-ship
   Decoy ship [ASW]

R

R^3
   Requirements Resources Review Board (DOD)
RA
   (1) Random Access Memory
   (2) Regular Army [slang]
   (3) Remedial Action [Environmental Protection Agency]
   (4) Reserve Affairs
RAAF
   Royal Australian Air Force
RAC
   Request for Authority to Complete
RACC
   Regional ASW Command Center (USN)
RACER
   Rankine-Cycle Energy Recovery System (USN)
Rad
   (1) Radar
   (2) Radio

RADALT
　Radar Altitude (USN)
radar**
　Radio Detection And Ranging
RADC
　Rome Air Development Center [N.Y.] (USAF)
RADINT
　Radar Intelligence
RADS
　Rapid Area Distribution Support (USAF)
RAF
　Royal Air Force [British]
RAG
　Replacement Air Group [later Readiness Air Group (USN); no longer used except as slang]
RAID
　Reconnaissance And Interdiction Detachment (USA)
RAKE
　Rocket Assisted Kinetic Energy (USA)
RAM
　(1) Radar-Absorbing Material
　(2) Reconnaissance Air Meet
　(3) Reliability, Availability, and Maintainability (USA)
　(4) Rolling Airframe Missile [missile designation RIM-116]
　(5) U.S. prefix for Soviet-Russian developmental aircraft flown at the Ramenskoye test facility
RAMP
　Rapid Acquisition of Manufactured Parts (USN)
RAMTIP
　Reliability and Maintainability Technology Insertion Program (USAF)
RAN
　(1) Request for Authority to Negotiate
　(2) Royal Australian Navy
RAP
　(1) Resource Allocation Process
　(2) Rocket Assisted Projectile
RAPID
　Real-time Automated Personnel System
RAPS
　Resource Analysis and Planning System (DOD)

RASL
　Reserve Active Status List
RASP
　Rapid Acquisition of Spare Parts (USA)
RAST
　Recovery Assistance, Securing, and Traversing System (USN)
RAT
　Ram Air Turbine (USN)
RATD
　Russian Aviation Trade House [Russia; established in 1991]
RATO
　Rocket Assisted Take-Off
RATT
　Radio Teletypewriter
RAV
　Restricted Availability (USN)
RAWS
　Radar Altimeter Warning System
RBM
　Readiness Based Maintenance (USA)
RC
　Reserve Component
RCAF
　Royal Canadian Air Force
RCAS
　Reserve Component Automation System (USA-USN)
RCC
　(1) Regional Control Center (USAF)
　(2) Rescue Coordination Center (USAF)
RCCM
　Regional Contingency Construction Management
RCHB
　Reserve Cargo-Handling Battalion (USN)
RCM
　Requirements Correlation Matrix (USAF)
RCP
　(1) Reenlistment Control Point
　(2) Retention Control Point (USA)
RCPAC
　Army Reserve Personnel and Administrative Center

RCPE
 Radiological Control Practice Evaluation (USN)
RCRA
 Resource Conservation and Recovery Act
RCS
 Radar Cross Section
RCT
 Regimental-sized Combined Arms Team (USMC)
RCTDAP
 Reserve Components Training Development Action Plan (USA)
RCU
 Remote Control Unit (USA)
RD
 Restricted Data [classification]
R&D
 Research and Development
RDA
 Research, Development, and Acquisition (USA)
RDEC
 Research, Development, and Engineering Center (USA)
RDF
 (1) Radio Direction Finding
 (2) Rapid Deployment Force [incorporated into CENTCOM]
RDIT
 Rapid Deployment Imagery Terminal
RDIXS
 Research and Development Information Exchange System (USN)
RDJTF
 Rapid Deployment Joint Task Force [now part of CENTCOM]
RDSS
 Rapidly Deployable Surveillance System (USN)
RDT&E
 Research, Development, Test, and Evaluation
REACT
 Rapid Execution And Combat Targeting (USAF)
READIEX
 Readiness Exercise (USN)
REC
 (1) Radio-Electronic Combat [Soviet-Russian]
 (2) Regional Evaluation Center (USN)
RECCE
 Reconnaissance

RECO
Remote Control [of mines]
Recon
Reconnaissance
REDCOM
Readiness Command
REDCON
Readiness Condition
RED HORSE
Rapid Engineering Deployable, Heavy Operational Repair Squadron, Engineer (USAF)
REF
Risk Evaluation Force
REFORGER
Return of Forces to Germany
REFTRA
Refresher Training
reg**
(1) regulation [slang]
(2) regular [slang for serviceman]
Regt
Regiment
(Rein)
Reinforced
REINF
Reinforced
REMIS
Reliability and Maintainability Information System (USAF)
REMS
Remotely Employed Sensors
REO
Regional Environmental Offices (USAF)
Re/Re
Reinforcement/Resupply [to Europe]
RES
Reserve
RESCAP
Rescue Combat Air Patrol
RESCOMMIS
Naval Reserve Command Management Information System
RESCORT
Rescue Escort (USN)

RESFORON
 Reserve Force Squadron (USN)
RF
 Radio Frequency
RFA
 Royal Fleet Auxiliary [British]
RFB
 Request For Bid
RFC
 Request for Comment
RFI
 (1) Radio Frequency Interference
 (2) Ready For Issue
 (3) Request For Information
RFI/EMI
 Radio Frequency Interference/Electromagnetic Interference
Rfl
 Rifle
RFP
 Request For Proposal
RFPB
 Reserve Forces Policy Board (DOD)
RFQ
 Request For Quotation
RHAWS
 Radar Homing And Warning System
RHIB
 Rigid-Hull Inflatable Boat
RHIP
 Rank Has Its Privileges [slang]
RI
 Remedial Investigation [Environmental Protection Agency]
RIB
 Rubberized Inflatable Boat
RIBS
 Readiness In Base Service (USAF)
RIF
 Reduction In Force
RIMPAC
 Pacific Rim
RIMS
 Revised Intertheater Mobility Study

RINT
　Radiation Intelligence
RIO
　Radar Intercept Officer (USN)
RIP
　Readiness Improvement Program (USN)
RIT
　Remote Imagery Transceiver
RIW
　Reliability Improvement Warranty
RJ
　Rivet Joint [RC-135 reconnaissance aircraft] (USAF)
Rkt
　Rocket
RL
　Rocket Launcher
R/L
　Receive Location
RLD
　(1) Ready-to-Load Date [at ORG]
　(2) Remote Launch Demonstration (USA)
RLG
　Ring Laser Gyro
RLSO
　Regional Logistical Support Offices
RLT
　Regimental Landing Team (USMC)
RM
　(1) Radioman
　(2) Royal Marines [British]
R&M
　Reliability and Maintainability (USAF)
RMP
　Reprogrammable Microprocessor
RMS
　Rocket Management System (USA)
RN
　Royal Navy [British]
RNLN
　Royal Netherlands Navy
RNMCB
　Reserve Naval Mobile Construction Battalion

RNZAF
Royal New Zealand Air Force
RNZN
Royal New Zealand Navy
ROA
Reserve Officers Association
ROB
Radar Order of Battle
ROC
(1) Republic of China [Taiwan]
(2) Required Operational Capability (USA-USMC)
ROCC
Region Operations Control Center (USAF)
ROD
Record Of Decision
ROE
Rules Of Engagement
ROEX
Rules Of Engagement Exercise
RO/FLO
Roll-On/Float-Off
ROH
Regular Overhaul (USN)
ROI
Return On Investment
ROK
Republic Of Korea [South]
ROKAF
Republic Of Korea Air Force
ROM
(1) Read-Only Memory
(2) Refuel-On-the-Move (USA)
RON
(1) Squadron [suffix; such as, PHIBRON, DESRON, and CORTRON]
(2) Remain Over-Night
ROPMA
Reserve Officer Personnel Management Act
ROR
Rapid-Onset-Rate (USAF)
RO/RO
Roll-On/Roll-Off (USN)

RORSAT
 Radar Ocean Reconnaissance Satellite
ROS
 Reduced Operating Status (USN)
ROTC
 Reserve Officer Training Corps
R-OTH
 Relocatable Over-The-Horizon [radar]
ROV
 Remote Operating Vehicle
ROW
 Rest Of World
ROWPU
 Reverse Osmosis Water Purification Unit
RP
 (1) Rally Point
 (2) Release Point
RPH
 Remotely Piloted Helicopter
RPI
 Roller Path Inclination (USN)
RPM
 Real Property Maintenance
RPMA
 Real Property Maintenance Activity (USA)
RPV
 (1) Remotely Piloted Vehicle
 (2) Reserve Personnel Navy
R&R
 Rest and Recuperation
RRA
 Ready Reserve Agreement (USN)
RRAD
 Red River Army Depot
RRC
 (1) Regional Reporting Centers (USN)
 (2) Rigid Raiding Craft
RRF
 Ready Reserve Force (USN)
RRFWG
 Ready Reserve Force Working Group

RSI
  Rationalization, Standardization, and Interoperability
RSIP
  Radar System Improvement Program
RSO
  Regional Security Officer
RSR
  Required Supply Rate (USA)
RSRA
  Rotor System Research Aircraft [X-wing] (NASA)
RSS
  Rosette Scan Seeker (USA)
RSSC
  Regional Signals Intelligence Support Center (NSA)
RSTA
  Reconnaissance, Surveillance, and Target Acquisition (USA)
RTASS
  Remote Tactical Airborne SIGINT System (USAF)
RTB
  Return(ed) To Base
RTC
  Reserve Training Center (USA)
RTO
  Responsible Test Organization
RTR
  Recruit Training Regiment (USMC)
RTSV
  Real-Time Synthetic Video
RTU
  Reserve Training Unit (USAF)
RU
  Intelligence Directorate [Soviet-Russian designation]
RUM
  Resource and Unit Monitoring
RV
  Reentry Vehicle
RWR
  Radar Warning Receiver

S

S1
  Staff Officer for Personnel (USA-USMC)
S2
  Staff Officer for Intelligence (USA-USMC)
S3
  Staff Officer for Operations (USA-USMC)
S4
  Staff Officer for Logistics (USA-USMC)
(S)
  Secret
SA
  (1) Secretary of the Army
  (2) Security Assistance
  (3) Selective Availability
  (4) Situational Awareness (USN)
  (5) Systems Analysis
SA-( )*
  Designation for Soviet-Russian Surface-to-Air missile (NATO)
SAAF
  South African Air Force
SAAM
  Special Assignment Airlift Mission
SAAWC
  Sector Anti-Air Warfare Coordinator (USMC)
SABM
  Systems Analysis/Battle Management (USA)
SABMIS
  Sea-based Anti-Ballistic Missile Intercept System [discarded concept]
SAC
  (1) Senate Appropriations Committee
  (2) Standard Aircraft Characteristics
  (3) Strategic Air Command [deactivated 1 June 1992] (USAF)
  (4) Supreme Allied Commander
SAC(A)
  Supporting Arms Coordinator (Airborne) (USMC)

SACC
 (1) Shore ASW Command Center
 (2) Supporting Arms Coordination Center
SACCS
 SAC Automated Command and Control System
SACEUR
 Supreme Allied Commander Europe (NATO)
SACEX
 Supporting Arms Coordination Exercise
SACLANT
 Supreme Allied Commander Atlantic (NATO)
SACOS
 SAC Operations Staff (USAF)
SACPMC
 Systems Acquisition Career Management Program for Civilians (USAF)
SADARM
 Sense And Destroy Armor [munition]
SADBUS
 Small And Disadvantaged Business Utilization Specialist
SADM
 System Acquisition Decision Memorandum (USA)
SADR
 Secure Acoustic Data Relay
SAE
 Service Acquisition Executive
SAF
 (1) Secretary of the Air Force
 (2) Soviet Air Forces [plural]
 (3) Special Action Force
SAFE
 Safe Areas For Evasion
SAG
 (1) Senior Advisory Group (USA)
 (2) Study Advisory Group (USA)
 (3) Surface Action Group [succeeded by Maritime Action Group] (USN)
SAGA
 Studies, Analysis, and Gaming Agency
SAGE
 Semiautomatic Ground Environment [obsolete] (USAF)

SAIE
 Special Acceptance and Inspection Equipment
SAL
 Semiactive Laser
SALFAS
 Stand-Alone Low-Frequency Active Sonar
SALT
 Strategic Arms Limitation Talks
SALTS
 Streamlined Alternative Logistics Transmission System
SAM
 (1) Semiautonomous Acoustic/Magnetic vehicle
 (2) Special Airlift Mission
 (3) Surface-to-Air Missile
 (4) Systems Acquisition Management
SAMS
 School of Advanced Military Studies (USA)
SAMSO
 Space And Missile Systems Organization (USAF)
SAMTEC
 Space And Missile Test and Evaluation Center (USAF)
SAMTO
 Space And Missile Test Organization (USAF)
SA-N-( )*
 Designation for Soviet-Russian Surface-to-Air missile, Naval (NATO)
SAO
 Security Assistance Organizations
SAP
 (1) Semi-Armor Piercing [munition]
 (2) Special Access Program
SAPRWG
 Security Assistance Program Review Working Group
SAR
 (1) Safety Assessment Report
 (2) Search And Rescue
 (3) Selected Acquisition Report
 (4) Special Access Required
 (5) Subsequent Application Review
 (6) Synthetic Aperture Radar

( )SARC
　Service System Acquisition Review Council; ( ) = A for Army, N for Navy, or AF for Air Force
SAREX
　Search And Rescue Exercise (USN)
SARS
　Standardized Army Refueling System
SARSAT
　Search And Rescue Satellite Aided Tracking (USAF)
SART
　Strategic Aircraft Reconstitution Team (USAF)
SARTS
　Small Arms Readiness Training Section (USA)
SAS
　(1) Special Air Service [Royal Air Force]
　(2) Storage Aids Systems (USAF)
SASC
　Senate Armed Services Committee
SASS
　Small Aerostat Surveillance System (USA)
SASSY
　Support Activities Supply System (USMC)
Sat
　Satellite
SATCOM
　Satellite Communications
SATKA
　Surveillance, Acquisition, Tracking, and Kill Assessment
SATMO
　Security Assistance Training Management Officer
SAU
　Squadron Augmentation Unit (USN)
SAUV
　Semiautonomous Underwater Vehicle
SAV
　Strike-Attack Vectoring (USN)
SAW
　(1) Squad Assault Weapon (USMC)
　(2) Squad Automatic Weapon (USA)
SAWS
　Submarine Acoustic Warfare System (USN)

SBA
Small Business Administration
SBC
Senate Budget Committoo
SBI
(1) Space-Based Interceptor
(2) Special Background Investigation
SBIR
Small Business Innovation Research Program
SBP
Survivor Benefit Plan
SBS
(1) Shipboard Simulators (USN)
(2) Special Boat Squadron (USN)
(3) Special Boat Squadron [Royal Marines]
SBSS
Standard Base Supply System
SBU
Small Boat Unit
SBWAS
Space-Based Wide-Area Surveillance (USAF)
SCA
Shipbuilders Council of America
SCAR
Strike Control and Reconnaissance (USAF)
SCB
Ship Characteristics Board [1945–1966]
SCBCA
Small Claims Board of Contract Appeals
SCCB
Software Configuration Control Board
SCDL
Surveillance Control Data Link
SCE
Standard Communication Environment
SCEPS
Stored Chemical Energy Propulsion System [torpedo]
SCFRS
Surface Combatant Force Requirements Study (USN)
SCG
Special Consultative Group

SCI
  (1) Sensitive Compartmented Information
  (2) Special Compartmented Intelligence
SCIB
  Ship Characteristics Improvement Board (USN)
SCINS
  Self-Contained Inertial Navigation System
SCMP
  Software Configuration Management Plan
SCN
  (1) Shipbuilding and Conversion, Navy [funding]
  (2) Software Change Notice
  (3) Specification Change Notice
SCNS
  Self-Contained Navigation System
SCOTT
  Single Channel Objective Tactical Terminal
SCP
  (1) Secure Conferencing Project
  (2) Strategic Computing Program [DARPA]
  (3) System Concept Paper [obsolete]
SCPE
  Simplified Chemical Protective Equipment (USA)
SCPS
  Survivable Collective Protection System (USAF)
SCSC
  Strategic Conventional Standoff Capability (USAF)
scuba**
  Self-Contained Underwater Breathing Apparatus
SD
  Space Division (USAF)
SDAF
  Special Defense Acquisition Fund (USA)
SDB
  (1) Small Disadvantaged Business Program
  (2) Standard Dress Blue (USN)
SDBUP
  Small Disadvantaged Business Utilization Program
SDC
  (1) Shaft-Driven Compressor
  (2) Situation Display Console

SDDM
  Secretary of Defense Decision Memorandum
SDF
  Standard Distance File
SDI
  Strategic Defense Initiative ["Star Wars"]
SDIO
  Strategic Defense Initiative Organization [renamed Ballistic Missile Defense Organization—BMDO—in 1993] (DOD)
SDL
  (1) Software Development Laboratory
  (2) Software Development Library
SDLS
  Satellite Data Link Standard
SDO
  Squadron/Staff Duty Officer
SDP
  (1) Software Development Plan
  (2) Standard Distance Package
SDR
  (1) Software Design Review
  (2) System Design Review
SDS
  (1) Self-Defense Suite (USAF)
  (2) Strategic Defense System
  (3) Surveillance Direction System
SDT
  Self-Development Test (USA)
SDV
  SEAL Delivery Vehicle [formally Swimmer Delivery Vehicle]
SDVT
  SEAL Delivery Vehicle Team
SE
  Systems Engineering
SEA
  (1) Sea Echelon Areas (USN)
  (2) Southeast Asia
SEABEE
  Sea Barge
SEACOP
  Strategic Sealift Contingency Planning System (MSC)

SEAD
　　Suppression of Enemy Air Defenses
SEAL
　　Sea, Air, Land (USN)
SEAM
　　Sidewinder Expanded-Acquisition Mode
SEAOPS
　　Safe Engineering and Operations Program (USMC)
SEASAT
　　Sea Satellite
SEATO
　　Southeast Asian Treaty Organization [defunct]
SECDEF
　　Secretary of Defense
SECNAV
　　Secretary of the Navy
SECNAVINST
　　Secretary of the Navy Instruction
SECTRANS
　　Secretary of Transportation
SEF
　　Stability Enhancement Function (USAF)
SEI
　　Space Exploration Initiative
SELRES
　　Selected Reserve (USA)
SEM
　　(1) Standard Equipment Modules (USN)
　　(2) Systems Engineering Management
SE/M
　　Systems Engineering/Management
SEMA
　　Special Electronic Mission Aircraft (USA)
SEMATECH
　　Semiconductor Manufacturing Technology
SEMP
　　System Engineering Management Plan
SENSO
　　Sensor Operator (USN)
SEP
　　(1) Selective Employment Plan
　　(2) Separate (USA)

(3) Spherical Error Probable
(4) System Engineering Process
SER
(1) Safety Evaluation Report
(2) Selective Early Retirement
SERB
Selective Early Retirement Board (USAF)
SERD
Support Equipment Requirements Document
SERDP
Strategic Environment Research and Development Program
SERE
Survival, Evasion, Resistance, and Escape [school] (USN)
SERP
Strategic Environmental Research Program
SERT
Shipboard Electronic Repair Team (USN)
SERV
Service
SES
(1) Senior Executive Service [civilian employee]
(2) Surface Effect Ship
SETA
Systems Engineering and Technical Assistance
SEW
Space and Electronic Warfare
SEWC
Space and Electronic Warfare Commander (USN)
SEWS
Satellite Early Warning System (USAF)
SF
Special Forces
SFARP
Strike Fighter Advanced Readiness Program (USN)
SFDLR
Stock Funding of Depot-Level Repairable (USA)
SFG
Special Forces Group (USA)
SFOD
Special Forces Operational Detachment
SFTS
Synthetic Flight Training System (USA)

SFW
 Sensor-Fused Weapon
SGLI
 (1) Service Government Life Insurance
 (2) Servicemen's Group Life Insurance
SH
 Station Hospital (USN)
SHAMP
 Ship Acquisition Project Manager (USN)
SHAPE
 Supreme Headquarters, Allied Powers, Europe (NATO) [formerly Supreme Headquarters Allied *Personnel*, Europe]
SHBD INT
 Basic Shipboard Intelligence
SHF
 Super High Frequency
SHORAD
 Short-Range Air Defense
SHORAD C^2
 Short-Range Air Defense Command and Control
SI
 (1) Signals Intelligence
 (2) Special Intelligence [classification]
S&I
 Survey and Investigations [organizations of the House Appropriations Committee]
SIA
 Station of Initial Assignment (USMC)
SIC
 (1) Senior Intelligence Committee
 (2) Standard Industrial Classification
 (3) Subscriber Interface Control
SICBM
 Small Intercontinental Ballistic Missile
SICC
 Service Item Control Centers (USA)
SICPS
 Standardized Integration Command Post System (USA)
SICR
 Specific Intelligence Collection Requirement
SIDS
 Secondary Imagery Dissemination System

SIG
　Senior Interdepartmental Group [NSC]
SIGINT
　Signals Intelligence
SIGIPS
　Signals Information Processing System (USN)
SIGSEC
　Signal Security
SIMA
　Ships Intermediate Maintenance Activity
SIMATS
　Supplemental Interim Medium Anti-Tank System (USA)
SIMNET
　Simulation Networking
SIMS
　Station Information Management System (USN)
SINCGARS
　(1) Single-Channel Ground-and-Air Radio System
　(2) Single-Channel Ground-and-Airborne Radio System
SINCGARS-V
　Single-Channel Ground-and-Air Radio System, VHF
SIO/EIA
　Ship Intelligence Officer/Enlisted Intel Assistant
SIOP
　Single Integrated Operational Plan
SIP
　Standardization Instructor Pilot (USA)
SIR
　Specific Information Requirement
SISMS
　Standard Integrated Support Management System
SITREP
　Situation Report
SITSUM
　Situation Summary
SJA
　Staff Judge Advocate
SKO
　Set, Kit, or Outfit
SKUL
　Seeker-Killer-Utility Lasers

SLA
Strategic Logistics Agency
SLAM
Standoff Land Attack Missile [modified Harpoon; missile designation AGM-84E]
SLAMMR
Sideways-Looking Airborne Multi-Mode Radar
SLAR
Side-Looking Airborne Radar [*not* aircraft]
SLAT
(1) Ship Launched Air-Targeted [missile]
(2) Strike Leader Attack Training (USN)
(3) Supersonic Low-Altitude Target [system; drone]
SLATS
Strike Leader Attack Training School (USN)
SLBD
Sea Lite Beam Director (USN)
SLBM
Submarine Launched Ballistic Missile
SLC
Submarine Laser Communications
SLCM
Sea-Launched Cruise Missile [formerly Submarine-Launched Cruise Missile]
SLEP
Service Life Extension Program (USN-USCG)
SLFCS
Survivable Low-Frequency Communications System (USAF)
SLGR
Small, Lightweight Global Positioning System Receiver
SLKT
Survivability, Lethality, and Key Technologies (USA)
SLMM
Submarine Launched Mobile Mine
SLOC(s)
Sea Line(s) Of Communication [not sea *lanes*]
SLOT
Submarine Launched One-way Transmitter [AN/BRT-1]
SLT
Squadron Landing Team (USMC)
SLUF
Short Little Ugly Feller [slang for A-7 Corsair] (USN)

SLV
 (1) Small Launch Vehicle (USAF)
 (2) Space Launch Vehicle
SM
 (1) Standard Missile
 (2) System Monitor
SMA
 Special Mission Aircraft
SMAW
 Shoulder-Launched Multi-purpose Assault Weapon (USA-USMC)
SMAW-D
 Shoulder-Launched Multi-purpose Assault Weapon—Disposable
SMCM
 Surface Mine Countermeasures (USN)
SMCR
 Selected Marine Corps Reserve (USMC)
SMD
 (1) Ship Manning Document (USN)
 (2) Standardized Military Drawing Program
SMDG
 Standoff Mine Detection Ground (USA)
SMDPS
 Strategic Mission Data Preparation System (USAF)
SME
 Squadron Medical Element (USAF)
SMES
 Superconductive Magnetic Energy Storage System
SMESA
 Special Middle East Sealift Agreement
SMG
 Submachine Gun
SMI
 Soldier-Machine Interface (USA)
SMIP
 Spares Management Improvement Program
SMOTEC
 Special Missions Operational Test and Evaluations Center (USAF)
SMP
 Soviet Military Power [publication, 1981-1991; changed in 1992 to *Forces in Transition*]

SMSF
 Special Mission Support Force (USN)
SMTP
 Simple Mail Transfer Protocol
SNA
 (1) Soviet Naval Aviation
 (2) Surface Navy Association
SNAFU
 Situation Normal, All F——d Up [slang]
SNAP
 (1) Shipboard Non-tactical Automatic Data Processing [program]
 (2) Systems for Nuclear Auxiliary Power [discarded concept]
SNCO
 Senior Non-Commissioned Officer
SNDL
 (1) Special Navy Distribution List
 (2) Standard Navy Distribution List
SNDM
 Secretary of the Navy Decision Memorandum
SNDV
 Strategic Nuclear Delivery Vehicle
SNEP
 Saudi Naval Expansion Program
SNF
 (1) Secret/No Foreign [distribution]
 (2) Short-range Nuclear Forces
SNI
 Soviet Naval Infantry
SNL
 Standard Nomenclature List
SNR
 Signal-to-Noise Ratio
SO
 Special Operations
SOA
 (1) Special Operations Aircraft (USA)
 (2) Speed of Advance (USN)
SOAC
 Submarine Officer Advanced Course (USN)
SOAR
 Special Operations Aviation Regiment

SOAS
Special Operations ADP System
SOB
(1) SAM Order of Battle
(2) Soviet Order of Battle
SOBC
Submarine Officer Basic Course (USN)
SOC
(1) Satellite Operations Complex (USAF)
(2) Sector Operations Center
(3) Special Operations Capability (USMC)
(4) Special Operations Command [see USSCOM]
SOCC
Submarine Operations Command Center
SOCCENT
Special Operations Command, CENTCOM
SOCCT
Special Operations Combat Control Team
SOCEUR
Special Operations Command, Europe
SOCEX
Special Operations Capable Exercise
SOC-PAC
Special Operations Command—Pacific (USA)
SOCRATES
Special Operations Command Research, Analysis, and Threat Evaluation System
SOCS
Ship Operational Characteristics Study
SOD
Strategy and Options Decision [PPBS]
SODS
Strategic Offensive Delivery Systems
SOE
Standard Option Equipment
SOF
Special Operations Force(s)
SOFA
Status Of Forces Agreement
SOG
Special Operations Group (USAF)

SOJ
Stand-Off Jamming
SOJS
Stand-Off Jammer Suppression
SOJT
Supervised On-the-Job Training
SO/LIC
Special Operations/Low-Intensity Conflict (USMC)
SOLL
Special Operations Low Level
SOLOG
Standardization of Operations and Logistics
SOMSS
Submarine Off-Board Mine Search System
SOMTE
Soldier Operator-Maintainer-Tester-Evaluator
SON
Statement of Operational Need (USAF)
sonar
Sound Navigation And Ranging [originally Sounding, Navigating, and Ranging]
SOP
Standard Operating Procedure [originally Standing Operating Procedure]
SOPA
Senior Officer Present Afloat (USN)
SOPC
Shuttle Operations and Planning Complex (USAF)
SOR
(1) Source of Repair (USAF)
(2) Special Operational Requirement
(3) Strategy and Options Review
SORD
System of Operational Requirements Document (USAF)
SORDAC
Special Operations Research and Development Center
SORM
Ships Organization Manual
SORN
Standard Organization and Regulations of the U.S. Navy
SORTS
Status Of Readiness and Training System

SOS
Squadron Officer School (USAF)
SOSB
(1) Special Operations Signal Battalion
(2) Special Operations Support Battalion
SOSS
Soviet [Russian] Ocean Surveillance System
SOSUS
Sound Surveillance System [seafloor]
SOT
Systems Operability Test
SOTA
Signals Intelligence Operational Tasking Authority
SOTG
Special Operations Training Group (USMC)
SOUTHAF
U.S. Air Force, U.S. Southern Command
SOW
(1) Stand-Off Weapon
(2) Statement Of Work (DOD)
SP
(1) Self-Propelled
(2) Shore Patrol (USN)
(3) Single-Purpose [gun]
SPACC
Space Control Center
SPACECOM
U.S. Space Command
SPAWAR
Space and Naval Warfare Systems Command (USN)
SPAWARSYSCOM
Space and Naval Warfare Systems Command (USN)
SPC
Statistical Process Control
SPCC
Ships Parts Control Center (USN)
SPEAR
(1) Special Project Evaluation and Anti-Air Warfare Research (USN)
(2) Strike Projection Evaluation and Anti-Air Warfare Research
SPEC
Specification

SPECAT
Special Category Message
SPECOPS
Special Operations (USN)
SPECWAR
Special Warfare
SPG
Special Planning Group [SOF]
SPICE
Space Integrated Controls Experiment
SPINS
Special Instruction
SPIREP
Spot Intelligence Report
SPLL
Self-Propelled Launcher Loader (USA)
SPM
(1) Software Programmer's Manual
(2) System Program Manager
SPMAGTF
Special Purpose Marine Air Ground Task Force
SPO
(1) Special Projects Office [now SSPO] (USN)
(2) System Program Office (USAF)
(3) System Project Office
SPOD
Sea Port Of Debarkation
SPOE
Sea Port Of Embarkation
SPR
(1) Secretarial Performance Review (DOD)
(2) Secretarial Program Review (USAF)
(3) Sponsor's Program Review (USN)
SPRAA
Strategic Plans and Resource Analysis Agency
SPS
(1) Simplified Processing Station
(2) Software Product Specification
SPSC
System Planning and System Control (USAF)
Sqdn
Squadron

SQEP
　Software Quality Evaluation Plan
SR
　(1) Special Reconnaissance [SOF]
　(2) Strategic Reconnaissance
SRA
　(1) Selected Reserve Augmentee
　(2) Selected Restricted Availability [in shipyard] (USN)
　(3) Shop Replaceable Assembly
　(4) Specialized Repair Activity
　(5) Special Repair Activity (USA)
SRAAM
　Short-Range Air-to-Air Missile
SRAM
　Short-Range Attack Missile [AGM-69A]
SRB
　(1) Selective Reenlistment Bonus (USN)
　(2) Service Record Book (USMC)
　(3) Sold Rocket Booster (NASA)
SRBM
　Short-Range Ballistic Missile
SRC
　(1) Strategic Reconnaissance Center (USAF)
　(2) Submarine Rescue Chamber [McCann rescue bell]
SRD
　Systems Requirement Document
SRF
　(1) Secure Reserve Force
　(2) Strategic Rocket Forces [Soviet-Russian]
　(3) Summary Reference File
SRFCS
　Self-Repairing Flight Control System (USAF)
SRI
　(1) Stanford Research Institute [Stanford, Calif.]
　(2) Surveillance, Reconnaissance, and Intelligence (USMC)
SRIG
　Surveillance, Reconnaissance, and Intelligence Group (USMC)
SRINF
　Shorter Range Intermediate-Range Nuclear Forces
SRM
　Solid Rocket Motor

SRP
(1) Sealift Readiness Program (USN)
(2) SIOP Reconnaissance Plan (USAF)
SRR
System Requirements Review
SRS
Software Requirements Specification
SRTC
Search Radar Terrain Clearance
SRT(s)
Strategic Relocatable Target(s)
SRU
(1) Shop Replacement Unit (USA)
(2) Subassembly Repairable Unit
(SS)
Submarine warfare [USN qualification]
SS-( )*
Designation for Soviet-Russian Surface-to-Surface missile (NATO)
SSA
(1) Security Supporting Assistance
(2) Software Support Activity
(3) Software Support Agency
(4) Source Selection Authority
(5) Special Support Activity [NSA]
(6) Supply Support Arrangement
SSAC
Source Selection Advisory Council
SSB
Special Separation Benefit
SSC
(1) Sea Surveillance and Coordination (USN)
(2) Senior Service College
(3) Strategic Systems Committee (DOD)
(4) Subspecialty Codes
(5) System/Site Control
SSC-( )*
Designation for Soviet-Russian Surface-to-Surface missile, Coastal (NATO)
SSC-N-( )*
Designation for Soviet-Russian Surface-to-Surface missile, Coastal, Naval (NATO)

SSCRA
Soldier and Sailor Civil Relief Act
SSD
Space System Division (USAF)
SSDS
Single Ship Deep Sweep
SSEB
Source Selection Evaluation Board
SSES
Ship Signals Exploitation Space
SSF
Special Security Force
SSG
(1) Ships Service Generator
(2) Special Study Group (USA)
(3) Strategic Studies Group [Naval War College]
SSIC
Standard Subject Identification Code
SSIXS
Submarine Satellite Information Exchange System (USN)
SSLV
Standard Small Launch Vehicle
SSM
Surface-to-Surface Missile
SS-N-( )*
Designation for Soviet-Russian Surface-to-Surface missile, Naval (NATO)
SSO
Special Security Office
SSORM
Standard Submarine Operations and Regulations Manual
SSP
(1) SIGINT Support Plan
(2) Source Selection Plan
SSPM
Software Standards and Procedures Manual
SSPO
Strategic Systems Projects Office (USN)
SSR
Software Specification Review
SSS
System/Segment Specification

SSSC
    Surface/Subsurface Surveillance Coordinator
SSTD
    Surface Ship Torpedo Defense (USN)
SSTO
    Single-Stage-To-Orbit
SSTS
    Space Surveillance and Tracking System
SSUFT
    Single Station Unit Fielding Training (USAF-USN)
SSWG
    System Safety Working Group
ST
    Special Tooling
S&T
    Science and Technology
STA
    Surveillance and Target Acquisition (USMC)
STAFS
    Supportable Technology for Affordable Fighter Structures (USAF)
STAMID
    Standoff Minefield Detection System (USA)
STAMIS
    Standard Army Management Information Systems
STANAG
    Standardization Agreement (NATO)
STANAVFORCHAN
    Standing Naval Force Channel (NATO)
STANAVFORLANT
    Standing Naval Force Atlantic (NATO)
STANAVFORMED
    Standing Naval Force Mediterranean (NATO)
STANG
    Standardization Agreement (NATO)
STAR
    (1) Scheduled Theater Airlift Route (USAF)
    (2) Stream Tension Actuated Remotely (USN)
    (3) Surface-To-Air Retrieval
    (4) System Threat Assessment Report
START
    Strategic Arms Reduction Talks [follow-on to SALT]

STD
 (1) Software Test Description
 (2) Standard
STE
 Special Test Equipment
STEP
 Stripes for Exceptional Performers (USAF)
STI
 Scientific Technical Information
STIC
 SEAL Tactical Insertion Craft (USN)
STIP
 Scientific Technical Information Program
STIR
 Separate Track and Illumination Radar (USN)
STK
 Strike (USN)
STLDD
 Software Top-Level Design Document
STO
 Short Take-Off
STOL
 Short Take-Off and Landing
STON
 Short Ton
STOVL
 Short Take-Off/Vertical Landing
STP
 (1) Software Test Plan
 (2) Submarine Technology Program [DARPA]
STPR
 Software Test Procedures
STR
 (1) Software Test Report
 (2) Strength
STRAT
 Strategic
STRATCOM
 Strategic Command [see USSTRATCOM]
STRATSAT
 Strategic Satellite (USAF)

STREAM
   Standard Tensioned Replenishment Alongside Method (USN)
STRIKEFORSOUTH
   Striking and Support Forces Southern Europe (NATO)
STROG
   Strait Of Gibraltar
STT
   SEAL Tactical Training
STTP
   Space Test and Transportation Program
STU
   Secure Telephone Unit
STUFT
   Ships Taken Up From Trade [British]
STW
   Strike Warfare (USN)
STWC
   Strike Warfare Commander (USN)
STWO
   Staff Tactical Watch Officer
STX
   Situational Training Exercise
Su
   Sukhoi [Soviet-Russian aircraft designation]
SUB
   Submarine
SUBACS
   Submarine Advanced Combat System [now AN/BSY-1 and AN/BSY-2]
SUBMIS
   Submarine Missing [signal]
SUBNOTE
   Submarine Notice (USN)
SUBOPAUTH
   Submarine Operating Authority (USN)
SUBROC
   Submarine Rocket [discarded]
SUBSUNK
   Submarine Sunk [emergency signal]
SUCAP
   Surface Combat Air Patrol

SUM
 (1) Shallow Underwater Mine
 (2) Software User's Manual
SUPCOM
 Support Command (USA)
SUPPLOT
 Supporting Plot
SUPSHIP
 (1) Superintendent of Shipbuilding
 (2) Supervisor of Shipbuilding
SUPT
 Specialized Undergraduate Pilot Training (USAF)
SURF EWO
 Surface Electronic Warfare Officer Course
SUROBs
 Surf Observations
SURTASS
 Surveillance Towed-Array Sonar System
SUSV
 Small Unit Support Vehicle (USA)
SUW-N-( )*
 Designation for Soviet-Russian Surface-to-Underwater missile, Naval (NATO)
S/V
 Survivability/Vulnerability
SVC
 (1) Service (USA)
 (2) Special Verification Commission
SVIP
 Secure Voice Improvement Program
SVLA
 Steered Vertical Line Array [sonobuoy]
SVML
 Standard Vehicle-Mounted Launcher
SVR
 Shop Visit Rate
SVS
 Secure Voice System
SW
 (1) Shallow Water
 (2) Software

S/W
　Software
(SW)
　Surface Warfare [USN qualification]
SWA
　Southwest Asia
SWAPDOP
　Southwest Asia Petroleum Distribution Operational Project
SWATH
　Small Waterplane Area Twin Hull [ship]
SWATS
　Sea-Based Weapons and Advance Tactics School
SWDG
　Surface Warfare Development Group (USN)
SWIP
　Systems Weapon Improvement Program [A-6 Intruder]
SWMCM
　Shallow-Water Mine Countermeasures
SWO
　Surface Warfare Officer (USN)
SWORD
　Submarine Warfare Operations Research Department
SWOS
　Surface Warfare Officers School
SWPS
　Strategic War Planning System (USAF)
SWS
　Special Warfare Systems
SWT
　Scout Weapons Team (USA)
SY
　Shipyard [also SYd; U.S. usage is generally NSY for Naval Shipyard]
SYDP
　Six-Year Defense Plan [used briefly from the late 1980s to 1991]
SYERS
　Senior Year Electro-optical Reconnaissance System (USAF)
SYSCOM
　System Command
SYSTO
　Systems Officer
SZ
　Surf Zone

T

TA
Table of Allowances
TAA
(1) Tactical Assembly Area (USA)
(2) Total Army Analysis
TAACOM
Theater Area Army Command
TAAF
Test, Analyze, And Fix
TABS
Total Army Basing Study
TAC
(1) Tactical
(2) Tactical Air Command [abolished 1992] (USAF)
TAC(A)
(1) Tactical Air Control (Airborne) (USMC)
(2) Tactical Air Coordinator Airborne (USAF)
TACAIR
Tactical Air [aircraft]
TACAMO
Take Charge And Move Out [communications relay aircraft program] (USN)
TACAN
Tactical Aircraft Navigation [system]
TACC
Tactical Air Command Center (USMC)
TACCIMS
Theater Automated Command, Control, and Information Management System
TACCO
Tactical Coordinator
TACCS
Tactical Army CSS Computer System (USA)
TACDEP
One-Day Tactical Deception Orientation
TACELINT
Tactical Electronic Intelligence (USN)
TACINTEL
Tactical Intelligence

TACJAM
Tactical Jammer Advanced Countermeasures System Upgrade (USA)
TACLET
Tactical Law Enforcement Team (USCG)
TACMEMO
Tactical Memo
TACMS
Tactical Missile System
TACOM
Tank Automotive Command (USA)
TACON
Tactical Control
TACP
Tactical Air Control Party
TACREP
Tactical Report
TACRON
(1) Tactical Air Control Squadron (USMC)
(2) Tactical Air Squadron
TACS
Tactical Air Control System
TACSAT
Tactical Satellite
TACSOP
Tactical SOP (USA)
TACSS
Tactical Army Combat Service Support
TACTAS
Tactical Towed-Array Sonar
TACTASS
Tactical Towed-Array Surveillance System
TACTRAGRU
Tactical Training Group (USN)
TAD
(1) Technology Area Description
(2) Temporary Additional Duty
TADIL
Tactical Digital Information Link
TADIXS
Tactical Data Information Exchange System (USN)

TADS
  (1) Target Acquisition and Data System (USA)
  (2) Target Acquisition and Designation Sight (USA)
TADSIXS-B
  Tactical Data Information Exchange System-B
TADSS
  Training Aids, Devices, Simulators, and Simulations (USA)
TAF
  Tactical Air Force
TAFHG
  Tactical Air Force Headquarters (USAF)
TAFIG
  Tactical Air Forces Interoperability Group (USAF)
TAFT
  Training Assistance Field Team (USA)
TAG
  The Adjutant General (USA)
TAGCEN
  The Adjutant General Center (USA)
TAGO
  The Adjutant General's Office (USA)
TAI
  Target Area of Interest
TALC
  Tactical Airborne Laser Communication
TALD
  Tactical Air Launched Decoy
TALM
  Tomahawk Land-Attack Missile [missile designation BGM-109]
TALO
  Tactical Airlift Liaison Officer (USAF)
TAMMIS
  Theater Army Medical Management and Information System
TAMP
  (1) Terminally Guided Anti-Armor Mortar Projectile (USN)
  (2) Theater Aviation Maintenance Program (USA)
TAMPS
  Tactical Aircraft Mission Planning System
TAO
  (1) Tactical Action Officer (USN)
  (2) Transition Assistance Office (USA)

TAOC
Tactical Air Operations Center (USMC)
TAOM
Tactical Air Operations Module (USMC)
TAPC
Total Army Personnel Command
TAR
(1) Temporary Active Reserve (USN)
(2) Training and Administration of Reserve (USN)
TARCAP
Target Combat Air Patrol (USN)
TARN
Tactical Air Request Net (USAF)
TARPS
Tactical Air Reconnaissance Pod System (USN)
TARS
Tactical Air Reconnaissance System
TAS
(1) Target Acquisition System (USA)
(2) True Airspeed
TASCFORM
Technique for Assessing Comparative Force Modernization
TASDAC
Tactical Secure Data Communication (USAF)
TASM
Tomahawk Anti-Ship Missile [missile designation BGM-109]
TASOSC
Theater Army Special Operations Support Command
TASS
Towed-Array Surveillance System
TAT
Technical Assistance Team (USA)
TAV
(1) Technical Availability (USN)
(2) Transatmospheric Vehicle
TAWP
Tactical Air Working Party
TBD
(1) To Be Determined
(2) To Be Developed
TBIP
Tomahawk Baseline Improvement Program

TBL
Turbulent-Boundary Layer
TBM
(1) Tactical Ballistic Missile
(2) Theater Ballistic Missile
TBMD
(1) Tactical Ballistic Missile Defense
(2) Theater Ballistic Missile Defense
TBS
The Basic School (USMC)
TBTC
Transportable Blood Transshipment Center
TC
(1) Training Circular (USA)
(2) Transportation Corps (USA)
(3) Type Classified (USA)
TCAC
Technical Control and Analysis Center
TC ACCIS
Transportation Coordination Automated Command and Control Information System (USA)
TCAE
Technical Control and Analysis Element
TC-AIMS
Transportation Coordinator's Automated Information for Movement System
TCAT
Type Commander Amphibious Training
TCATC
TRADOC Combined Army Test Activity (USA)
TCC
(1) Tactical Command Center
(2) Transportation Component Command
TCCT
Type Commander Core Training
TCO
(1) Tactical Combat Operations
(2) Termination Contracting Officer
TCP
Tactical Computer Processor (USA)
TCPS
Transportable Collective Protection System (USAF)

TCS
  (1) Television Camera System (USN)
  (2) Total Commissioned Service
TCT
  Tactical Computer Terminal (USA)
TD
  (1) Table of Distribution
  (2) Technical Data
  (3) Technical Director
  (4) Test Director
TDA
  (1) Tables of Distribution and Allowance
  (2) Tactical Decision Aid
TDAC
  Training Data and Analysis Center
TDAP
  Training Development Action Plan
TDBM
  Tactical Data Base Manager
TDD
  Target Detection Device
TDF
  Tactical Digital Facsimile
TDG
  Tactical Decision Game (USMC)
TDP
  (1) Tactical Data Processor
  (2) Technical Data Package
  (3) Technology Development Plan (USA)
  (4) Test Design Plan
TDRSS
  Tracking and Data Relay Satellite System
TDS
  Tactical Data System
TDY
  Temporary Duty
T/E
  Table of Equipment
TE
  Test Equipment
T&E
  Test and Evaluation

TEAMS
Tactical EA-6 Mission Planning System (USN)
TEC
Total Estimated Cost
TECHEVAL
Technical Evaluation
TECHMOD
Technology Modernization
TECOM
Test and Evaluation Command (USA)
TEECG
Tactical Exercise Evaluation Control Group (USMC)
TEL
Transporter, Erector, and Launcher [for missiles]
TELNET
Telecommunications Network
TEMP
Test and Evaluation Master Plan
TEMPER
Tent Extendable Modular Personnel (USA)
TEMSE
Technical and Managerial Support Environment
TENCAP
Tactical Exploitation of National Capabilities
TERA
Temporary Early Retirement Authority
TERCOM
Terrain Contour Matching [guidance]
TEREC
Tactical Electronic Reconnaissance (USAF)
TERPES
Tactical Electronic Reconnaissance Processing and Evaluation System
TERPROM
Terrain Profile Matching
TERPS
Tactical Electronic Reconnaissance Processing and Evaluation System (USN)
TERS
Tactical Event Reporting System
TEWS
Tactical Electronic Warfare System

TEXCOM
Test and Experimentation Command (USA)
TEXS
Tactical Explosive System (USA)
TF
Task Force
TFCC
(1) Tactical Flag Command Center
(2) Task Force Command Center (USN)
TFE
(1) Tactical Field Exchange (USAF)
(2) Transportation Feasibility Estimator
TFU
Tactical Forecast Unit (USAF)
TG
Task Group
TGIF
Tactical Ground Intercept Facility (USAF)
TGS
Turreted Gun System (USA)
TGSM
Terminally Guided Submunitions
TGW
Terminal Guidance Warhead
THAAD
Theater High-Altitude Area Defense (USA)
THMT
Tactical High-Mobility Terminal
TI
(1) Texas Instruments, Inc.
(2) Total Inventory
TIAP
Theater Intelligence Architecture Program
TIARA
Tactical Intelligence and Related Activities [program]
TIBS
Tactical Information Broadcast System (USAF)
TIC
Tactical Information Coordinator
TIG
Time In Grade
TILO
Technical Industrial Liaison Office (USA)

TIM
Technical Interchange Meeting
TIMS
TFCC Information Management System
TINS
Thermal Imaging Navigation Set
TIO
Target Information Officer (USMC)
TIP
Technical Information Pilot
TIRS
Target Index Reference System (USA)
TIS
(1) Thermal Imaging System
(2) Time in Service
(3) Tracking Instrumentation Subsystem (USAF)
TISEO
Target Identification System Electro-Optical
TIWG
Test Integration Working Group (USA)
TJAG
The Judge Advocate General (USA)
Tk
Tank
TK
Talent Keyhole [classification]
TLA
Temporary Lodging Allowance
TLAM
Tomahawk Land-Attack Missile
TLAM-C
Tomahawk Land-Attack Missile (Conventional)
TLAM-N
Tomahawk Land-Attack Missile (Nuclear)
TLCF
Teleconference [WIN]
TLE
(1) Temporary Lodging Entitlement
(2) Temporary Lodging Expense
(3) Treaty Limited Equipment
TLI
Treaty Limited Item

TLS
Time Line Sheet
TM
(1) Technical Management
(2) Technical Manual
TMD
(1) Tactical Missile Defense (USA)
(2) Theater Missile Defense (USA)
TMDE
Test, Measurement, and Diagnostic Equipment
TMDI
(1) Tactical Missile Defense Initiative
(2) Theater Missile Defense Initiative (USA)
TMIS
Theater Medical Information System
TMMM
Tomahawk Multi-Mission Missile
TNF
Theater Nuclear Force[s]
TNMCS
Total Not-Mission Capable, Supply (USAF)
TNT
Trinitrotoluene
TO
(1) Table of Organization
(2) Technical Order
T/O
(1) Table of Organization
(2) Take Off
TOA
(1) Table of Allowance
(2) Total Obligational Authority [budget]
TOC
Tactical Operations Center (USA)
TOE
Tables of Organization and Equipment
TO&E
Tables of Organization and Equipment
TOSS
Tactical Operations Support System
TOT
Time On Target

TOW
    Tube-launched, Optically tracked, Wire-guided missile [missile designation MGM-71]

TP^2
    Transport Protocol Class 2

TP
    (1) Technical Performance
    (2) Transaction Processing

TPC
    Total Project Cost

TPCSDS-T
    Target Practice Cone Stabilized Discarding Sabot with Tracer (USA)

TPFDD
    Time-Phased Force and Deployment Data

TPFDL
    Time-Phased Force and Deployment List

TPM
    Technical Performance Measurement

TPS
    (1) Test Package Set
    (2) Test Pilots School (USN)
    (3) Test Program Set (USA)
    (4) Time Phase Line

TP-T
    Target Practice with Tracer (USA)

TPU
    Troop Program Unit

TPW
    Target Planning Worksheet

TPWG
    Test Planning Working Group (USAF)

TQG
    Tactical Quiet Generator (USA)

TQL
    Total Quality Leadership

TQM
    Total Quality Management

T/R
    Transmit/Receive

TR
    (1) Tactical Reconnaissance
    (2) Technical Report

(3) Theater Reserve (USA)
(4) Travel Requests
T&R
Training and Readiness (USMC)
TRA
(1) Taiwan Relations Act [1979]
(2) Training
TRAC
Tactical Radar Correlator (USA)
TRACALS
Traffic Control, Approach, and Landing System (USAF)
TRACE
Total Risk Assessing Cost Estimating
TRADOC
Training and Doctrine Command (USA)
TRAM
(1) Target Recognition and Attack Multisensor (USN)
(2) Tractor, Rubber-Tired, Articulated, Multipurpose
TRANSCOM
Transportation Command [see USTRANSCOM]
TRANSLANT
Transit Atlantic [by ship or aircraft]
TRANSPAC
Transit Pacific [by ship or aircraft]
TRAP
Tactical Receive Equipment and Related Applications
TRASANA
TRADOC Systems Analysis Activities (USA)
TRC
Technology Repair Center (USAF)
TRE
(1) Tactical Readiness Evaluation [submarines]
(2) Tactical Receive Element
(3) Tactical Receive Equipment
TRI-TAC
Joint Tactical Communications Program
Triad
U.S. strategic offensive forces [*not* an acronym]
TRICOMS
Triad Computer Systems
TRIGS
TR-1 [aircraft] Ground Station (USAF)

TRIMIS
    Tri-Service Medical Information System
TROSCOM
    Troop Support Command (USA)
TRP
    Target Reference Point (USA-USMC)
TRR
    Test Readiness Review
TRS
    Tactical Reconnaissance System
TRSA
    Training System Requirements Analysis
TRT
    Tanker Recovery Team (USAF)
TRUE
    Training in Urban Environment (USN)
TRV/SRV
    Tower Restoral Vehicle and Surveillance Restoral Vehicle (USAF)
TS
    Top Secret [classification]
TSG
    The Surgeon General
TSIP
    TOW Sight Improvement Program (USA)
TSIR
    Total System Integration Responsibility
TSM
    TRADOC System Manager
TSPI
    Time and Space Positioning Information
TSPR
    Total System Performance Responsibility
TSS
    (1) Tactical Shelter System
    (2) Tactical Surveillance Sonobuoy
    (3) Total Ship Survivability
TSSAM
    Tri-Service Stand-off Attack Missile [missile designation AGM-137]
TSTA
    Tailored Ship Training Availability (USN)
TSU
    Telescopic Sighting Unit (USA)

TT
(1) Teletype
(2) Torpedo Tube
TTAD
Temporary Tour of Active Duty
TTCC
Tomahawk Tactical Commanders Course
TTE
Technical Training Equipment
TT&E
Technical Test and Evaluation (USA)
TTF&T
Technology Transfer, Fabrication, and Test
TTS
Tank Thermal Sight (USA)
TTT
Time To Target (USA)
TTTS
Tanker/Transport Training System
TTU
Terminal Transportation Unit (USA)
TTY
(1) Teletype
(2) Teletypewriter
TTZ
Performance and design specification [Soviet-Russian[5]]
Tu
Tupelov [Soviet-Russian aircraft designation]
TU
Task Unit
TUCHA
Type Unit Data File
TUDET
Type Unit Equipment Detail File
TUSA
Third United States Army
TVA
Target Value Analysis

---

[5] *Taktiko-tekhnicheskoe zadanie.*

TVC
　Thrust Vectoring Control [system]
TVD
　Theater of Military Operations [Soviet-Russian]
TW&AA
　Tactical Warning and Attack Assessment (USAF)
TWCS
　Tomahawk Weapon Control System
TWMP
　Track Width Mine Plow (USMC)
TWOS
　Total Warrant Officer System
TWS
　(1) Thermal Weapon Sight (USA)
　(2) Track-While-Scan [radar]
TWV
　Tactical Wheeled Vehicle
TYCOM
　Type Commander (USN)
TYT
　Type Training

U

(U)
　Unclassified
UARS
　Unmanned Aerial Reconnaissance System
UARV
　Unmanned Air Reconnaissance Vehicle
UAV
　Unmanned Aerial Vehicle
UBA
　Underwater Breathing Apparatus [British]
UBRD
　Usage Based Requirements Determination (USA)

UCA
Undefinitized Contractual Actions
UCFF
UTC Consumption Factors File
UCI
Updated Coordinating Instructions
UCMJ
Uniform Code of Military Justice
U-COFT
Unit Conduct Of Fire Trainer (USA)
UCP
Unified Command Plan
UCR
Unit Cost Report
UDF
Unit Development Folder
UDL
Unit Designation List
UDT
Underwater Demolition Team [no longer used]
UE
Unit Equipment
UEP
Underwater Electrical Potential
UER
Unit Equipment Report (USMC)
UFO
UHF Follow On
UH
Utility Helicopter (USA)
UHF
Ultra-High Frequency
UI
Unit of Issue
UIC
Unit Identification Code
ULC
Unit Level Code
ULCANS
Ultra-Lightweight Camouflage Net System
ULCC
Ultra-Large Crude Carrier [ship]

ULCS
 Unit-Level Circuit Switch
ULMS
 Underwater Long range Missile System [now Trident SLBM]
ULN
 Unit Line Number
UMA
 Unit Mobilization Augmentation (USA)
UMMIPS
 Uniform Material Movement and Issue Priority System
UMTE
 Unmanned Threat Emitter (USAF)
UN
 United Nations
UNAAF
 Unified Action Armed Forces
UNC
 United Nations Command
UNCLASS
 Unclassified
UNFICYP
 United Nations Forces In Cyprus
UNHCR
 United Nations High Commission for Refugees
UNIFIL
 United Nations Interim Force In Lebanon
UNIKOM
 United Nations–Iraq–Kuwait Observer Mission
UNISAMS
 Universal Naval Integrated Surface-to-Air Missile System
UNK
 Unknown(s)
UNOLS
 University National Oceanographic Laboratory System (USN)
UNREP
 Underway Replenishment (USN)
UNSC
 United Nations Security Council
UNSECNAV
 Under Secretary of the Navy
UNTSO
 United Nations Truce Supervision Organization [Middle East]

UPS
Uniform Procurement System
UPT
Undergraduate Pilot Training (USAF)
URL
Unrestricted Line [Officer]
U/S
Unified and Specified [Commands]
USA
(1) Under Secretary of the Army
(2) U.S. Army
USAAA
U.S. Army Audit Agency
USACAP
U.S. Army Chemical Activity, Pacific
USACC
U.S. Army Communications Command
USACDEC
U.S. Army Combat Development Experimentation Center
USACE
U.S. Army Corps of Engineers
USACFSC
U.S. Army Community and Family Support Center
USACIC
U.S. Army Criminal Investigation Command
USAEDS
U.S. Atomic Energy Detection System
USAEUR
U.S. Army Europe
USAF
U.S. Air Force
USAFA
U.S. Air Force Academy
USAFAC
U.S. Army Finance and Accounting Center
USAFAGOS
U.S. Air Force Air Group Operations School
USAFALCENT
U.S. Air Force Airlift Center
USAFE
U.S. Air Forces in Europe
USAFR
U.S. Air Force Reserve

USAFSAM
U.S. Air Force School of Aerospace Medicine
USAFSNCOA
U.S. Air Force Senior Noncommissioned Officer Academy
USAFSO
U.S. Air Force Southern Air Division
USAFSOC
U.S. Air Force Special Operations School
USAHPSA
U.S. Army Health Professional Support Agency
USAHSC
U.S. Army Health Services Command
USAIA
U.S. Army Intelligence Agency
USAIRCENT
U.S. Air Forces, U.S. Central Command
USAISC
U.S. Army Information Systems Command
USAKA
U.S. Army Kwajalein Atoll
USAMMA
U.S. Army Medical Material Agency
USAMMCE
U.S. Army Medical Material Center—Europe
USAMMCSA
U.S. Army Medical Material Center—Saudi Arabia
USAPPC
U.S. Army Publications and Printing Command
USAR
U.S. Army Reserve
USARC
U.S. Army Reserve Command
USARCENT
U.S. Army Forces, U.S. Central Command
USAREC
U.S. Army Recruiting Command
USAREUR
U.S. Army, Europe
USARJ
U.S. Army, Japan
USARPAC
U.S. Army, Pacific

USARSO
U.S. Army South
USARSPACE
U.S. Army Space Command
USASA
U.S. Army Security Agency
USASAALA
U.S. Army Security Assistance Agency, Latin America
USASAC
U.S. Army Security Assistance Command
USASCH
U.S. Army Support Command, Hawaii
USASDC
U.S. Army Strategic Defense Command
USASG
U.S. Army Support Group
USASOC
U.S. Army Special Operations Command
USATSA
U.S. Army Troop Support Agency
USATSC
U.S. Army Training Support Center
USAWC
U.S. Army War College
USBRO
U.S. Based Requirements Overseas
USC
U.S. Code
USCENTCOM
U.S. Central Command
USCG
U.S. Coast Guard
USCGC
U.S. Coast Guard Cutter
USCGR
U.S. Coast Guard Reserve
USC&GS
U.S. Coast and Geodetic Survey
USCINC
U.S. Commander in Chief
USCINCAFRED
U.S. Commander in Chief, Air Force Readiness Command

USCINCCENT
    Commander in Chief, U.S. Central Command
USCINCEUR
    Commander in Chief, U.S. European Command
USCINCFOR
    Commander in Chief, U.S. Forces Command
USCINCLANT
    Commander in Chief, U.S. Atlantic Command [changed to CINCUSACOM in 1993]
USCINCPAC
    Commander in Chief, U.S. Pacific Command
USCINCRED
    Commander in Chief, U.S. Readiness Command
USCINCSO
    Commander in Chief, U.S. Southern Command
USCINCSOC
    Commander in Chief, Special Operations Command
USCINCSOCOM
    Commander in Chief, U.S. Special Operations Command
USCINCSOUTH
    Commander in Chief, U.S. Southern Command
USCINCSPACE
    Commander in Chief, U.S. Space Command
USCINCTRANSCOM
    Commander in Chief, U.S. Transportation Command
USCINSTRAT
    Commander in Chief, Strategic Command
USD
    Under Secretary of Defense
USDA
    (1) U.S. Department of Agriculture
    (2) USD for Acquisition
USDAO
    U.S. Defense Attaché Office
USDP
    USD for Policy
USDR&E
    USD for Research and Engineering
USDRO
    U.S. Defense Representative Office
USEUCOM
    U.S. European Command [unified]

USFJ
U.S. Forces Japan
USFK
U.S. Forces Korea
USFORSCOM
U.S. Forces Command [specified]
USG
U.S. Government
USGS
U.S. Geological Survey
USHBP
Uniformed Service Health Benefit Program
USIA
(1) U.S. Information Agency
(2) U.S. Inspection Agency
USLANTCOM
U.S. Atlantic Command [unified]
USLO
U.S. Liaison Office
USLOK
U.S. Liaison Office—Kuwait
USLOT
U.S. Liaison Office—Tunisia
USMA
(1) U.S. Maritime Academy
(2) U.S. Maritime Administration [now MarAd]
(3) U.S. Military Academy
USMARCENT
U.S. Marine Corps, U.S. Central Command
USMC
U.S. Marine Corps
USMCMG
U.S. Mine Countermeasures Group
USMCR
U.S. Marine Corps Reserve
USML
U.S. Munitions List
USMMA
U.S. Merchant Marine Academy
USMTF
U.S. Message Text Format
USMTM
U.S. Military Training Mission

USN
  U.S. Navy
USNA
  U.S. Naval Academy
USNAVCENT
  U.S. Naval Forces, U.S. Central Command
USNI
  U.S. Naval Institute
USNIP
  U.S. Naval Institute *Proceedings* [magazine]
USNR
  U.S. Naval Reserve
USNS
  U.S. Naval Ship [MSC operated]
USO
  United Services Organization
USPACOM
  U.S. Pacific Command
USPS
  U.S. Postal Service
USR
  Unit Status Report
USREDCOM
  U.S. Readiness Command
USS
  U.S. Ship
USSOC
  U.S. Special Operations Command
USSOCOM
  U.S. Special Operations Command
USSOUTHCOM
  U.S. Southern Command
USSPACECOM
  U.S. Space Command
USSR
  Union of Soviet Socialist Republics [became CIS in 1992]
USSS
  U.S. SIGINT System
USSTRATCOM
  U.S. Strategic Command [established 1992]
USTF
  Uniformed Services Treatment Facilities

USTRANSCOM
U.S. Transportation Command
USUHS
Uniformed Services University of Health Sciences
UTA
Unity Training Assembly
UTARS
Utility Aircraft Requirements Study (USA)
UTC
Unit Type Code
UT&E
User Test and Evaluation (USA)
UUT
Unit Under Test (USA)
UUV
(1) Unmanned Underwater Vehicle
(2) Untethered Underwater Vehicle
UW
Unconventional Warfare

VA
(1) Department of Veterans Affairs
(2) Veterans Administration [now Department of Veterans Affairs]
(3) Veterans Affairs
VAOSC
Visibility and Management of Operations and Support Costs
VBSS
Visit, Board, Search, and Secure
VC
Variable Cost
VCJCS
Vice Chairman, Joint Chiefs of Staff
VCNO
Vice Chief of Naval Operations

VCSA
Vice Chief of Staff, Army
VCSAF
Vice Chief of Staff, Air Force
VDD
Version Description Document
VDL
Video Data Link
VDS
Variable Depth Sonar
VE
Value Engineering
VEAP
Veterans Education [Assistance Act of 1984]
VECP
Value Engineering Change Proposal
VENS
Versatile Exercise Mine System
VERRP
Voluntary Early Retirement Program
VERTREP
Vertical Replenishment (USN)
VFR
Visual Flight Rules
VGK
Supreme High Command [Soviet-Russian]
VGLI
Veteran's Group Life Insurance
VHA
Variable Housing Allowance
VHF
Very High Frequency [see Table 2 under EHF]
VHSIC
Very High Speed Integrated Circuit
VID
Visual Identification
VIDS
Visual Information Display System (USN)
VIDS/MAF
Visual Information Display System/Maintenance Action Form (USN)

VIP
  (1) Very Important Person [slang]
  (2) Visual Information Projection
VIS
  Vehicle Intercommunication System (USA)
VISTA
  Variable Stability In-flight Simulator Test Aircraft (USAF)
VLA
  Vertical Launch ASROC (USN)
VLAD
  Vertical Line Array DIFAR [sonobuoy]
VLCC
  Very Large Crude Carrier [tanker]
VLF
  Very Low Frequency
VLS
  Vertical Launch System (USN)
VLSI
  Very Large Scale Integration
VMC
  Visual Meteorological Condition (USN)
VMD
  Veterinary Medical Doctor
VOA
  Voice Of America
VOD
  Vertical On-board Delivery [VERTREP is preferred]
VOQ
  Visiting Officer Quarters
VOR VHF
  Omni-direction Range [navigation system]
VPK
  Military Industrial Commission [Soviet-Russian]
VPU
  Patrol Special Projects Unit (USN)
VSI
  Voluntary Separation Incentive
VSTOL
  Vertical/Short Take-Off and Landing
VSW
  Very Shallow Water

VT
(1) Proximity Fuse [Variable-Time]
(2) Virtual Terminal
VTA
Transport Aviation [Soviet-Russian]
VTAS
Visual Target-Acquisition System
VTO
Vertical Take-Off
VTOL
Vertical Take-Off and Landing
VTS
Vessel Traffic System (USCG)

WAA
Wide Aperture Array [sonar]
WAG
Wild-Assed Guess [slang]
WAL
Warfare Analysis Laboratory [Johns Hopkins University/Applied Physics Laboratory]
WAM
WWMCCS ADP Modernization
WAMC
Wide Area Mine Clearance (USA)
WAPS
Weighted Airman Promotion System
WARMAPS
Wartime Manpower Planning System
WASEX
War At Sea Exercise (USN)
WATCHCON
Watch Condition
WBGT
Wet Bulb Globe Temperature

WBS
　Work Breakdown Structure
WCC
　Warfare Commanders Course
WCO
　Weapons Control Officer (USN)
WDC
　Weapons Delivery Computer
WESTCOM
　U.S. Army Western Command
WESTPAC
　Western Pacific (USN)
WEU
　(1) Western Economic Union
　(2) Western European Union
WHNS
　Wartime Host Nation Support
WHNSIMS
　Wartime Host Nation Support Information Management System
WHS
　Washington Headquarters Services
WIA
　Wounded In Action
WIG
　Wing-In-Ground [effect vehicle]
WIN
　WWMCCS Intercomputer Network
WIPP
　Waste Isolation Pilot Plant
WIS
　WWMCCS Information System
WITS
　WSGT Intelligent Terminal System
wl
　waterline [length]
WMA
　Warfare Mission Area
WMP
　War and Mobilization Plan (USAF)
WO
　Warrant Officer
WOEC
　Warrant Officer Entry Course (USA)

WOS
　Warrant Officer Service (USA)
WOSC
　Warrant Officer Senior Course (USA)
WP
　(1) Warsaw Pact
　(2) Weapons Procurement
　(3) White Phosphorus
WPC
　(1) Warrior Preparation Center
　(2) World Peace Council
WPCC
　Wright-Patterson Contracting Center (USAF)
WPI
　Wholesale Price Index
WPN
　Weapons Procurement, Navy [funding]
WRM
　War Reserve Material (USAF)
WRS
　War Reserve Stocks
WRSA
　War Reserve Stock for Allies
WRSK
　War Readiness Spares Kit (USAF)
WSAM
　Weapon Systems Acquisition Management (USN)
WSEP
　Weapons System Evaluation Program
WSGT
　WWMCCS Standard Graphics Terminal
WSIG
　Weapons Support Improvement Group (DOD)
WSIP
　Weapons System Improvement Program
WSMIS
　Weapons System Management Information System (USAF)
WSMP
　Weapons System Master Plan (USAF)
WSMR
　White Sands Missile Range (USA)
WST
　Weapons System Trainer

WTT
  Warfare Commanders Team Training
WTVD
  Western Theater of Operations
WWABNCP
  Worldwide Airborne National Command Post (USAF)
WWABNRES
  WWMCCS Airborne Resources
WWIMS
  Worldwide Indicators and Monitoring System
WWMCCS
  Worldwide Military Command and Control System

XBT
  Expendable Bathythermograph
Xmtl
  Experimental
XO
  Executive Officer
XTB
  Experimental Test Bed (USA)

Yak
  Yakovlev [Soviet-Russian aircraft designation]
YOYO
  You're On Your Own [slang]

CHAPTER 2

Aircraft Designations

The current U.S. military aircraft designation scheme—shown in Table 3—was adopted in 1962. It provides an indication of the basic aircraft type and the aircraft's sequence within that type; prefix and suffix letters provide additional details. But confusion persists as the old (pre-1962) and new schemes are mixed or written incorrectly. For example, the McDonnell Douglas F-4B Phantom is sometimes written incorrectly as F4B—which was a Boeing fighter of the 1920s. Similarly, the F4F Wildcat of World War II fame is often written incorrectly as F-4F, which is the U.S. designation used for F-4 Phantoms configured for West Germany.

Further, the U.S. military services do not follow the system as it was established. For example, in 1962 a new helicopter series was established, beginning with H-1 (formerly HU-1); that series reached only to H-6 before the series began adding to the abandoned Air Force numbers with H-54 and above. More severe violations were performed with fighter-series aircraft numbers above F-111 being assigned although a new series had begun with the F-1 and carried through to the F-23. But the subsequent "stealth" fighter (F-117) and other fighter designs violate the scheme.

Also, modifications to aircraft, which have in the past added a new suffix numeral or letter, have instead resulted in such confusing designations as the P-3C UPDATE III, EA-6B ICAP, and EP-3E AIRES II aircraft.

AIRCRAFT DESIGNATIONS

HISTORICAL

The unified scheme for U.S. military aircraft went into effect in October 1962 when all existing and new naval aircraft and all new Air Force planes were redesignated in a new, simplified series, almost all beginning with the series number one. The Navy-flown AD Skyraider became the first plane in the new attack series, the A-1. The Navy's TF Trader started the new cargo series as C-1; the FJ Fury became the F-1; the T2V Sea Star the T-1; and the UC-1 Otter the U-1.

There was no P-1 or S-1, because the new system picked up the Navy's P2V Neptune and S2F Tracker as the P-2 and S-2, respectively. The improved P3V Orion was the obvious candidate for P-3 and the P5M Marlin, the Navy's last combat flying boat, for P-5. The designation P-4 was used, albeit briefly, for the drone versions of the Privateer (the P4Y-2K, formerly PB4Y-2). The designations P-4 and P-6 are sometimes cited as having been reserved for the P4M Mercator and the P6M Seamaster. But the last of the combination piston-turbojet Mercators were gone and the turbojet Seamaster flying boat had been cancelled before the new system was established in 1962. The next patrol aircraft was supposed to be the P-7, which has been cancelled.

The 1962 system also introduced the mission designation of special electronic *E*-series aircraft. The first two planes were Navy, the WF-2 Tracer became the E-1B, and the W2F-1 Hawkeye the E-2A.

Variations of the previous Air Force *X* (for experimental) and *V* (for VSTOL) designations remained, but official records differ as to which aircraft were part of the old or new series. The Marine AV-8 Harrier is officially in the *V* series, but apparently the designation A-8 was avoided to reduce confusion.

In the *V* series, however, the Ryan "flying jeep" had already been designated XV-8. The latter program never took off, hence the *8* spot is firmly held by the successful Harrier series.

Another naval aircraft that contributes to confusion is the F/A-18 Hornet. This aircraft began life as the F-18 in the fighter series, having evolved from the YF-17 prototype. When the decision was made to configure the aircraft as a strike fighter, the designation was changed to F/A-18 (although the two-seat trainer version remained simply TF-18).

Planes that were used by both services, such as the Albatross seaplane (Navy UF), generally took on the existing Air Force nu-

AIRCRAFT DESIGNATIONS

merical designation (U-16, formerly SA-16). But the Phantom was recent enough to be given a new designation, the now-familiar F-4, and not the Air Force F-110.

Helicopters proved a more confusing issue because the Army had still another designation series before 1962 in addition to those of the Navy–Marine Corps and Air Force. The Sea Knight was the Navy HRB, whereas the Army called the helicopter HC-1A (HC for helicopter—cargo). This became the H-46 in the new scheme. The Army's HU-1 Iroquois (HU for helicopter—utility) started the new helicopter series as H-1 with most of the Army and Air Force designations merging to form the new *H* series. Navy helicopters were stuck in where there were gaps. The Kaman HU2K became the H-2 and the Sikorsky HSS-2 the H-3, but the Navy's HSS-1/HUS, which was similar to the Army–Air Force's H-34, took on that designation.

Further, after the new *H* series reached H-6, the military reverted to simply adding to the larger numerical series, that is, H-60 and H-65.

See Table 3 for the current aircraft designation scheme.

In this chapter aircraft are listed below by their series designators: thus, the EA-6B Prowler electronics aircraft is listed after the A-6E Intruder, from whose basic design the EA-6B was derived. Similarly, major C-130 Hercules variants are listed immediately following the basic C-130 entry. The AV-8B Harrier VSTOL attack aircraft is listed in the attack series.

Not all variants currently in service are listed.

ATTACK SERIES

TA-4
Skyhawk (Douglas): Training and utility aircraft; all attack variants (A-4) have been discarded; designated A4D before 1962. 1 turbojet engine. (USN)

A-6E
Intruder (Grumman): All-weather attack aircraft; designated A2F before 1962. 2 turbojet engines. (USMC-USN)

EA-6B
Prowler (Grumman): Tactical electronic jamming aircraft; based on A-6 Intruder. 2 turbojet engines. (USMC-USN)

Table 3. U.S. Aircraft Designations

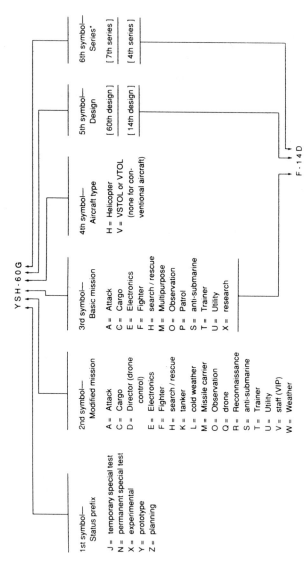

SOURCE: Norman Polmar, *The Naval Institute Guide to the Ships and Aircraft of the U.S. Fleet*, 15th ed. (Annapolis: Naval Institute Press, 1993), 397.

AIRCRAFT DESIGNATIONS

A-7D
 Corsair (LTV): Light attack aircraft; A-7K is a two-seat, combat-capable trainer; all Navy variants have been discarded. 1 turbojet engine. (USAF)

AV-8B
 Harrier (McDonnell Douglas–British Aerospace): VSTOL attack aircraft; also training variant (TAV-8B). 1 turbofan engine. (USMC)

A-10
 Thunderbolt (Fairchild Republic): Close air support and anti-tank aircraft; OA-10 used for FAC role. Also called "Warthog." 2 turbojet engines. (USAF)

OA-37
 Dragonfly (Cessna): Ground support aircraft; derived from the T-37. 2 turbojet engines. (USAF)

BOMBER SERIES

B-1B
 Lancer (North American Rockwell): Strategic bomber (variable-sweep wings). 4 turbojet engines. (USAF)

B-2A
 [no name] (Northrop): Strategic bomber (flying-wing; low-observable/stealth configuration); previously designated Advanced Technology Bomber (ATB). 4 turbojet engines. (USAF)

B-52
 Stratofortress (Boeing): Strategic bomber. 8 turbojet engines. (USAF)

CARGO SERIES

C-2A
 Greyhound (Grumman): Carrier On-board Delivery (COD). 2 turboprop engines. (USN)

C-4A
 Academe (Grumman): Executive transport. Also called "Gulfstream." 2 turboprop engines. (USCG)

TC-4C
 Academe (Grumman): Bombardier/navigator training aircraft for A-6E Intruder. 2 turboprop engines. (USN)

AIRCRAFT DESIGNATIONS

C-5
    Galaxy (Lockheed): Long-range, heavy transport. 4 turbofan engines. (USAF)

C-9
    Nightingale (McDonnell Douglas): Aeromedical evacuation and executive transport (VC-9) aircraft. 2 turbofan engines (USAF)

C-9
    Skytrain (McDonnell Douglas): Transport. 2 turbofan engines. (USN)

KC-10A
    Extender (McDonnell Douglas): Aerial tanker and transport aircraft. 3 turbofan engines. (USAF)

VC-11
    Gulfstream II (Grumman): Executive transport. 2 turbojet engines. (USCG)

C-12
    Huron (Beech): Transport. 2 turboprop engines. (USA-USAF-USMC-USN)

RC-12
    Huron (Beech): Signal intelligence and surveillance (Project Guardrail). 2 turboprop engines. (USA)

C-17
    Globemaster III (McDonnell Douglas): STOL transport. 4 turbofan engines. (USAF)

EC-18
    [no name] (Boeing): Advanced Range Instrumentation Aircraft (ARIA). 4 turbofan engines. (USAF)

C-20
    Gulfstream III (Gulfstream Aerospace): Executive transport. 2 turbofan engines. (USAF-USN)

C-21
    [no name] (Learjet): Transport. 2 turbofan engines. (USAF)

C-23
    Sherpa (Short Brothers Ltd.): Transport. 2 turboprop engines. (USAF)

EC-24
    [no name] (Douglas): Electronic Warfare (EW) simulation/jamming aircraft. 4 turbofan engines. (USN)

C-26A
    [no name] (Fairchild): Transport aircraft. 2 turboprop engines. (USAF)

AIRCRAFT DESIGNATIONS

C-29A
[no name] (British Aerospace): Air route inspection aircraft. 2 turbofan engines. (USAF)

C-130
Hercules (Lockheed): Cargo aircraft. Navy variants designated GV before 1962; LC-130 is Navy ski-equipped variant; MC-130 Combat Talon is USAF special operations variant. Also called "Herk." 4 turboprop engines. (USAF-USN)

AC-130
Spectre (Lockheed): Gunship. 4 turbojet engines. (USAF)

EC-130
Hercules and Compass Call (Lockheed): Various electronic configurations including ABCCC. (USAF)

EC-130V
Hercules (Lockheed): Surveillance aircraft. 4 turboprop engines. (USCG)

HC-130H
Hercules (Lockheed): Search-And-Rescue (SAR) aircraft. 4 turbofan engines. (USAF-USCG)

KC-130
Hercules (Lockheed): Aerial tanker and transport aircraft. 4 turbojet engines. (USN-USMC)

EC-135
Stratolifter (Boeing): Various electronic configurations including airborne command post. 4 turbofan engines. (USAF)

KC-135
Stratotanker (Boeing): Tanker aircraft. 4 turbojet engines. (USAF)

RC-135
Stratolifter (Boeing): Strategic reconnaissance aircraft. 4 turbojet engines. (USAF)

C-140
JetStar (Lockheed): Executive transport. 4 turbojet engines. (USAF)

C-141
Starlifter (Lockheed): Long-range cargo aircraft. 4 turbofan engines. (USAF)

ELECTRONIC SERIES

E-2C
Hawkeye (Grumman): Carrier-based Airborne Early Warning (AEW). 2 turbojet engines. (USN)

E-3
    Sentry (Boeing): Airborne Warning And Control System (AWACS); modified commercial 707. 4 turbofan engines. (USAF)

E-4
    [no name] (Boeing): Airborne National Command Post (ABNCP). 4 turbofan engines. (USAF)

E-6A
    Mercury (Boeing): Strategic communications aircraft (TACAMO); modified commercial 707. 4 turbofan engines. (USN)

E-8A
    [no name] (Boeing): Joint Surveillance/Target Attack Radar System (JSTARS) aircraft; modified commercial 707. 4 turbofan engines. (USA-USAF)

E-9A
    [no name] (Boeing Canada): Electronics test aircraft. 2 turboprop engines. (USAF)

FIGHTER SERIES

F-4
    Phantom (McDonnell Douglas): Fighter-attack aircraft; all Navy–Marine Corps variants have been discarded. 2 turbojet engines. (USAF)

F-5A/B
    Freedom Fighter (Northrop): Fighter-training aircraft. 2 turbojet engines. (USAF)

F-5E/F
    Tiger II (Northrop): Fighter-training aircraft. 2 turbojet engines. (USAF-USN)

F-14
    Tomcat (Grumman): Fighter aircraft; carrier based. 2 turbofan engines. (USN)

F-15
    Eagle (McDonnell Douglas): Fighter aircraft. 2 turbofan engines. (USAF)

F-15E
    Strike Eagle (McDonnell Douglas): Fighter-attack aircraft. 2 turbofan engines. (USAF)

F-16
    Fighting Falcon (Lockheed): Fighter-attack aircraft. 1 turbofan engine. (USAF)

AIRCRAFT DESIGNATIONS

F/A-18
Hornet (McDonnell Douglas–Northrop): Fighter-attack aircraft. 2 turbofan engines. (USN-USMC)

F-22
Lightning (Lockheed–General Dynamics–Boeing): Fighter aircraft; formerly Advanced Technology Fighter (ATF). 2 turbofan engines. (USAF)

F-111
Aardvark (General Dynamics): Fighter-attack aircraft (variable-sweep wings). 2 turbofan engines. (USAF)

EF-111
Raven (General Dynamics): Tactical electronic jamming aircraft; converted F-111. 2 turbofan engines. (USAF)

FB-111
[no name] (General Dynamics): Strategic bomber variant of F-111 (variable-sweep wings). 2 turbojet engines. (USAF)

F-117A
[no name] (Lockheed): All-weather strike aircraft (low-observable/stealth configuration). 2 turbofan engines. (USAF)

GLIDER SERIES

RG-8A
[no name] (Schweizer): Motorized surveillance glider. 1 piston engine. (USCG)

HELICOPTER SERIES

H-1
Iroquois (Bell): Transport; most are UH-1 with rescue (HH-1) and training (TH-1) variants also in service. Also called "Huey." 2 turboshaft engines. (USA-USAF-USMC-USN)

AH-1
Cobra (USA) and SeaCobra (USMC) (Bell): Gunship. 2 turboshaft engines. (USA-USMC)

SH-2
Seasprite (Kaman): Anti-submarine/Light Airborne Multipurpose System (LAMPS I). 2 turboshaft engines. (USN)

HH-3E
Jolly Green Giant (Sikorsky): Search-And-Rescue (SAR) helicopter. 2 turboshaft engines. (USAF)

AIRCRAFT DESIGNATIONS

SH-3
    Sea King (Sikorsky): Anti-submarine helicopter; also flown as presidential executive helicopter (VH-3). 2 turboshaft engines. (USMC-USN)

OH-6
    Cayuse (Hughes): Liaison helicopter. 1 turboshaft engine. (USA-USN)

H-46
    Sea Knight (Boeing Vertol): Transport (CH-46) and Vertical Replenishment (VERTREP/UH-46) helicopter. 2 turboshaft engines. (USMC-USN)

H-47
    Chinook (Boeing Vertol): Transport (CH-47) and special operations (MH-47) helicopter. 2 turboshaft engines. (USA)

H-53
    Super Jolly (USAF) and Sea Stallion (USMC-USN) (Sikorsky): Transport, rescue (HH-53), and special operations (MH-53 Pave Low) helicopter. 2 turboshaft engines. (USAF-USMC-USN)

CH-53E
    Super Stallion (Sikorsky): Transport helicopter. 3 turboshaft engines. (USMC-USN)

MH-53E
    Sea Dragon (Sikorsky): Mine Countermeasures (MCM) helicopter. 3 turboshaft engines (USN)

TH-57
    SeaRanger (Bell): Training helicopter. 1 turboshaft engine. (USN)

OH-58
    Kiowa (Bell): Observation and special operations (MH-58) helicopter. 1 turboshaft engine. (USA)

H-60
    Black Hawk and Night Hawk (Sikorsky): Transport helicopter; various configurations including electronics (EH-60), rescue (HH-60), special operations (MH-60 Pave Hawk), and executive transport (VH-60); Navy HH-60H is Seahawk and Coast Guard HH-60J is Jayhawk. 2 turboshaft engines. (USCG-USA-USAF-USMC-USN)

SH-60
    Sea Hawk (Sikorsky): Anti-submarine helicopter; the SH-60B is the LAMPS III. 2 turboshaft engines. (USN)

AH-64
    Apache (McDonnell Douglas–Hughes): Gunship. 2 turboshaft engines. (USA)

AIRCRAFT DESIGNATIONS

HH-65
 Dolphin (Aerospace Helicopter Corp. and Aérospatiale): Search-And-Rescue (SAR) helicopter. 2 turboshaft engines. (USCG)

RAH-66
 Comanche (Boeing-Sikorsky): Scout/light attack helicopter. 2 turboshaft engines. (USA)

OBSERVATION SERIES

OV-1/RV-1
 Mohawk (Grumman): Battlefield observation and reconnaissance aircraft. 2 turboprop engines. (USA)

PATROL SERIES

P-3
 Orion (Lockheed): Maritime patrol/anti-submarine aircraft. 4 turboprop engines. (USN)

RECONNAISSANCE SERIES

SR-71
 Blackbird (Lockheed): Strategic reconnaissance aircraft. 2 turbojet engines; all in reserve except for NASA-operated aircraft. (USAF)

TR-1
 [no name] (Lockheed): Tactical reconnaissance aircraft; all redesignated U-2. 1 turbojet engine. (USAF)

U-2
 [no name] (Lockheed): Strategic reconnaissance aircraft. 1 turbojet engine. (USAF)

ANTI-SUBMARINE SERIES

S-3
 Viking (Lockheed): Carrier-based anti-submarine aircraft. 2 turbojet engines. (USN)

ES-3A
 Viking (Lockheed): Carrier-based Signals Intelligence (SIGINT) aircraft. 2 turbojet engines. (USN)

AIRCRAFT DESIGNATIONS

TRAINER SERIES

T-2
> Buckeye (North American Rockwell): Undergraduate jet training aircraft; carrier capable; designated T2J before 1962. 2 turbojet engines. (USN)

T-34C
> Mentor (Beech): Primary training aircraft. 1 turboprop engine. (USN)

T-37
> Tweet (Cessna): Primary training aircraft. 2 turbojet engines. (USAF)

T-38
> Talon (Northrop): Advanced and aggressor training aircraft. 2 turbojet engines. (USAF-USN)

T-39N
> Sabreliner (Rockwell International): Advanced training aircraft. 2 turbojet engines. (USN)

CT-39
> Sabreliner (Rockwell International): Transport-utility aircraft. 2 turbojet engines. (USAF-USN)

T-41
> Mescalero (Cessna): Pilot candidate screening aircraft. 1 piston engine. (USAF)

T-43A
> [no name] (Boeing): Navigational training aircraft; modified commercial 737. 2 turbofan engines. (USAF)

T-44
> King Air (Beech): Multi-engine training aircraft. 2 turboprop engines. (USN)

T-45
> Goshawk (McDonnell Douglas–British Aerospace): Primary training aircraft. 1 turbofan engine. (USN)

T-47
> Citation (Cessna): Naval Flight Officer (NFO) training aircraft. 2 turbofan engines. (USN)

VSTOL SERIES

AV-8B
> Harrier (see Attack Series, above).

AIRCRAFT DESIGNATIONS

OV-10A
Bronco (Rockwell International): STOL Counterinsurgency (COIN) aircraft. 2 turboprop engines. (USAF)

XV-15A
[no name] (Bell): Tilt-rotor VSTOL technology demonstrator. 2 turboshaft engines. (NASA-USA-USAF-USN)

UV-18
Twin Otter (DeHavilland Canada): STOL transport. 2 turboprop engines. (USAF)

V-22
Osprey (Bell-Boeing Vertol): Multi-role tilt-rotor VSTOL aircraft; MV-22 is the proposed Marine troop carrier, EV-22 the proposed Navy AEW variant, HV-22 the proposed Navy SAR variant, SV-22 the proposed Navy ASW variant, and CV-22 the proposed Army special operations variant. 2 turboshaft engines. (USN-USMC)

EXPERIMENTAL SERIES

X-29
[no name] (Grumman): Forward-Swept Wing (FSW) demonstration aircraft. 1 turbofan engine. (NASA-USAF)

X-30
[no name] (undetermined): Planned National Aerospace Plane (NSP). (DOD-NASA)

X-31
[no name] (Rockwell-MBB): Enhanced Fighter Maneuverability (EFM) demonstrator. 1 turbojet engine. (NASA-USAF-USN)

CHAPTER 3

Aviation Unit Designations

AIR FORCE

AAG
  Aeromedical Airlift Group
AAS
  Aeromedical Airlift Squadron
ABG
  Air Base Group
AG
  Airlift Group
ALS
  Airlift Squadron
ARG
  Air Refueling Group
ARW
  Air Refueling Wing
ARS
  Air Refueling Squadron
AS
  Airlift Squadron
AW
  Airlift Wing
BMW
  Bombardment Wing
CAMS
  Consolidated Aircraft Maintenance Squadron

AVIATION UNIT DESIGNATIONS

CTG
    Crew Training Group
CTW
    Crew Training Wing
FG
    Fighter Group
FS
    Fighter Squadron
FTS
    Fighter Training Squadron
FW
    Fighter Wing
PBW
    Provisional Bombardment Wing
RQS
    Rescue Squadron
RQW
    Rescue Wing
SOG
    Special Operations Group
SOS
    Special Operations Squadron
SOW
    Special Operations Wing
SRS
    Strategic Reconnaissance Squadron
SRW
    Strategic Reconnaissance Wing
TAS
    Tactical Airlift Squadron
TAW
    Tactical Airlift Wing
TFS
    Tactical Fighter Squadron
TFW
    Tactical Fighter Wing
TRS
    Tactical Reconnaissance Squadron
WS
    Weather Squadron

AVIATION UNIT DESIGNATIONS

ARMY

ATKHB
Attack Helicopter Battalion

MARINE CORPS

Naval aviation units—Navy and Marine Corps—use the same system of abbreviations of simple letter-number combinations; however, the Navy additionally uses an acronym style for certain units, which is employed mainly in communications. Thus, Fighter Squadron 1 is known as both VF-1 and FITRON ONE while Attack Squadron 92 is both VA-92 and ATKRON NINETY-Two.

The *V* prefix for naval aircraft types and subsequently for aviation units dates from 1922, when *V* was used to indicate heavier-than-air and *Z* for lighter-than-air airships. *VF* indicates fighter squadron, *VA* attack squadron, *ZP* airship patrol squadron, and so on. (The last U.S. Navy airship, a Goodyear ZPG-2W, was taken out of service in 1962.) Subsequently, *H* was introduced as the helicopter type letter for aircraft (HNS-1) in 1943 and for squadrons in 1947 (the first squadron was Marine Corps HMX-1, followed in 1948 by Navy HU-1 and HU-2).[1]

Marine units have the letter *M* added as the second letter of aviation unit designations. Marine squadrons add the letter *T* as a suffix for readiness/transition units (such as VMAT).

HMA
Helicopter Attack Squadron
HMA/L
Helicopter Attack/Light Squadron
HMH
Helicopter Light Squadron
HMM
Helicopter Medium Squadron

---

[1] The *H*, *V*, and *Z* were also used for ship designations, hence CV for aircraft carrier, AV for seaplane tender, and AZ for airship tender. Subsequently, *H* was used in CHA, CVHE, LPH, LHA, and LHD for helicopter-carrying ships.

AVIATION UNIT DESIGNATIONS

HMX
 Helicopter Squadron [not Helicopter Development Squadron]
MACG
 Marine Air Control Group
MAG
 Marine Aircraft Group
MAW
 Marine Aircraft Wing
MAWTS
 Marine Aviation Weapons and Tactics Squadron
MCCRTG
 Marine Combat Crew Readiness Training Group
MWSS
 Marine Wing Support Squadron
VMA
 Attack Squadron
VMA(AW)
 Attack Squadron [All-Weather]
VMAQ
 Tactical Electronic Warfare Squadron
VMFA
 Fighter-Attack Squadron
VMFP
 Tactical Reconnaissance Squadron
VMGR
 Aerial Refueler-Transport Squadron
VMO
 Observation Squadron

NAVY

CVW
 Carrier Air Wing
CVWR
 Reserve Carrier Air Wing
FRS
 Fleet Readiness Squadron
HC
 Helicopter Combat Support Squadron
HCS
 Helicopter Combat Search and Rescue/Special Warfare Support Squadron

AVIATION UNIT DESIGNATIONS

HM
    Helicopter Mine Countermeasure Squadron
HS
    Helicopter Anti-Submarine Squadron
HSL
    Light Helicopter Anti-Submarine Squadron
HT
    Helicopter Training Squadron
RCVW
    Reserve Carrier Air Wing
VA
    Attack Squadron
VAK
    Aerial Refueling Squadron
VAQ
    Tactical Electronic Warfare Squadron
VAW
    Carrier Airborne Early Warning Squadron
VC
    Fleet Composite Squadron
VF
    Fighter Squadron
VFA
    Strike Fighter Squadron
VP
    Patrol Squadron
VPU
    Patrol Squadron—Special Projects Unit
VQ
    Fleet Air Reconnaissance Squadron[2]
VR
    Fleet Logistics Support Squadron
VRC
    Fleet Tactical Support Squadron [COD]
VS
    Air Anti-Submarine Squadron
VT
    Training Squadron

---

[2] VQ is also used for the TACAMO strategic communications relay squadrons.

AVIATION UNIT DESIGNATIONS

VX
 Air Test and Evaluation Squadron
VXE
 Antarctic Development Squadron
VXN
 Oceanographic Development Squadron

CHAPTER 4

Military Ranks

Grades indicate the pay level of personnel; the prefix O indicates officer, W indicates warrant officer, and E indicates enlisted.

AIR FORCE

		Grade
Gen	General	O-10
LtGen	Lieutenant General	O-9
MajGen	Major General	O-8
BrigGen	Brigadier General	O-7
Col	Colonel	O-6
LtCol	Lieutenant Colonel	O-5
Maj	Major	O-4
Capt	Captain	O-3
1stLt	First Lieutenant	O-2
2ndLt	Second Lieutenant	O-1
CMSgt	Chief Master Sergeant	E-9
SMSgt	Senior Master Sergeant	E-8
MSgt	Master Sergeant	E-7
TSgt	Technical Sergeant	E-6
SSgt	Staff Sergeant	E-5
Sgt	Sergeant	E-4
A1C	Airman First Class	E-3
AMN	Airman	E-2
AB	Airman Basic	E-1

MILITARY RANKS

Grade

ARMY

GEN	General	O-10
LTG	Lieutenant General	O-9
MG	Major General	O-8
BG	Brigadier General	O-7
COL	Colonel	O-6
LTC	Lieutenant Colonel	O-5
MAJ	Major	O-4
CPT	Captain	O-3
1LT	First Lieutenant	O-2
2LT	Second Lieutenant	O-1
CW5	Chief Warrant Officer 5	W-5
CW4	Chief Warrant Officer 4	W-4
CW3	Chief Warrant Officer 3	W-3
CW2	Chief Warrant Officer 2	W-2
CW1	Chief Warrant Officer 1	W-1
CMA	Command Sergeant Major of the Army	E-9
CSM	Command Sergeant Major	E-9
SGM	Sergeant Major	E-9
1SG	First Sergeant	E-8
MSG	Master Sergeant	E-8
PSG	Platoon Sergeant	E-7
SFC	Sergeant First Class	E-7
SSG	Staff Sergeant	E-6
SP6	Specialist 6	E-6
SGT	Sergeant	E-5
SP5	Specialist 5	E-5
CPL	Corporal	E-4
SP4	Specialist 4	E-4
PFC	Private First Class	E-3
PVT	Private	E-2
PVT	Private	E-1

MARINE CORPS

Gen	General	O-10
LtGen	Lieutenant General	O-9

MILITARY RANKS

		Grade
MajGen	Major General	O-8
BGen	Brigadier General	O-7
Col	Colonel	O-6
LtCol	Lieutenant Colonel	O-5
Maj	Major	O-4
Capt	Captain	O-3
1st LT	First Lieutenant	O-2
2nd LT	Second Lieutenant	O-1
CWO-5	Chief Warrant Officer 5	W-5
CWO-4	Chief Warrant Officer 4	W-4
CWO-3	Chief Warrant Officer 3	W-3
CWO-2	Chief Warrant Officer 2	W-2
CWO-1	Chief Warrant Officer 1	W-1
Sgt Maj	Sergeant Major	E-9
Mgy Sgt	Master Gunnery Sergeant	E-9
1st Sgt	First Sergeant	E-8
MSgt	Master Sergeant	E-8
GySgt	Gunnery Sergeant	E-7
SSgt	Staff Sergeant	E-6
Sgt	Sergeant	E-5
Cpl	Corporal	E-4
LCpl	Lance Corporal	E-3
PFC	Private First Class	E-2
Pvt	Private	E-1

NAVY AND COAST GUARD

ADM	Admiral	O-10
VADM	Vice Admiral	O-9
RADM	Rear Admiral[1]	O-8

[1]The Navy's Rear Admiral rank contains two levels: the upper half is equal to a two-star flag officer and the lower half is equal to a one-star flag officer. The rank of Commodore was a one-star rank during time of war. The commander of a unit of ships who is below the rank of rear admiral is sometimes referred to as *Commodore* as a courtesy title, regardless of his rank. The rank no longer exists in the U.S. Navy and the title is not officially used.

MILITARY RANKS

		Grade
RADM	Rear Admiral	O-7
CAPT	Captain	O-6
CDR	Commander	O-5
LCDR	Lieutenant Commander	O-4
LT	Lieutenant	O-3
LTJG	Lieutenant Junior Grade	O-2
ENS	Ensign	O-1
CWO5	Chief Warrant Officer	W-5
CWO4	Chief Warrant Officer	W-4
CWO3	Chief Warrant Officer	W-3
CWO2	Chief Warrant Officer	W-2
MCPO	Master Chief Petty Officer	E-9
SCPO	Senior Chief Petty Officer	E-8
CPO	Chief Petty Officer	E-7
PO1	Petty Officer 1	E-6
PO2	Petty Officer 2	E-5
PO3	Petty Officer 3	E-4
SN	Seaman	E-3
SA	Seaman Apprentice	E-2
SR	Seaman Recruit	E-1

NAVY SPECIALTY RATINGS

Navy enlisted ranks are identified by both pay grade and specialty; thus, a seaman striking as a quartermaster is a QMSN, whereas an aviation boatswain's mate second class is AB2.

AB
   Aviation Boatswain's Mate
ABE
   Aviation Boatswain's Mate (launching and recovering equipment)
ABF
   Aviation Boatswain's Mate (fuels)
ABH
   Aviation Boatswain's Mate (aircraft handling)
AC
   Air Traffic Controller

MILITARY RANKS

AD
    Aviation Machinist's Mate
AE
    Aviation Electrician's Mate
AG
    Aerographer's Mate
AK
    Aviation Storekeeper
AM
    Aviation Structural Mechanic
AME
    Aviation Structural Mechanic (safety equipment)
AMH
    Aviation Structural Mechanic (hydraulics)
AMS
    Aviation Structural Mechanic (structures)
AO
    Aviation Ordnanceman
AQ
    Aviation Fire Control Technician
AS
    Aviation Support Equipment Technician
ASE
    Aviation Support Equipment Technician (electrical)
ASM
    Aviation Support Equipment Technician (mechanical)
AT
    Aviation Electronics Technician
AW
    Aviation Anti-submarine Warfare Operator
AX
    Aviation Anti-submarine Warfare Technician
AZ
    Aviation Maintenance Administrationman
BM
    Boatswain's Mate
BT
    Boiler Technician
BU
    Builder
CE
    Construction Electrician

MILITARY RANKS

CM
    Construction Mechanic
CT
    Cryptologic Technician
CTA
    Cryptologic Technician (administrative)
CTI
    Cryptologic Technician (interpretive)
CTM
    Cryptologic Technician (maintenance)
CTO
    Cryptologic Technician (communications)
CTR
    Cryptologic Technician (collection)
CTT
    Cryptologic Technician (technical)
DC
    Damage Controlman
DK
    Disbursing Clerk
DM
    Illustrator Draftsman
DP
    Data Processing Technician
DS
    Data Systems Technician
DT
    Dental Technician
EA
    Engineering Aide
EM
    Electrician's Mate
EN
    Engineman
EO
    Equipment Operator
ET
    Electronics Technician
EW
    Electronic Warfare Technician
FC
    Fire Controlman

MILITARY RANKS

FT
    Fire Control Technician
FTB
    Fire Control Technician (ballistic missile fire control)
FTG
    Fire Control Technician (gunfire control)
FTM
    Fire Control Technician (missile fire control)
GM
    Gunner's Mate
GMG
    Gunner's Mate (guns)
GMM
    Gunner's Mate (missiles)
GMT
    Gunner's Mate (torpedoes)
GS
    Gas Turbine Systems Technician
GSE
    Gas Turbine Systems Technician (electrical)
GSM
    Gas Turbine Systems Technician (mechanical)
HM
    Hospital Corpsman
HT
    Hull Maintenance Technician
IC
    Interior Communications Electrician
IM
    Instrumentman
IS
    Intelligence Specialist
JO
    Journalist
LI
    Lithographer
LN
    Legalman
MA
    Master At Arms
ML
    Molder

MILITARY RANKS

MM
    Machinist's Mate
MN
    Mineman
MR
    Machinery Repairman
MS
    Mess Management Specialist
MT
    Missile Technician
MU
    Musician
NC
    Navy Counselor
OM
    Opticalman
OS
    Operations Specialist
OT
    Ocean Systems Technician
PC
    Postal Clerk
PH
    Photographer's Mate
PM
    Patternmaker
PN
    Personnelman
PR
    Aircrew Survival Equipmentman
QM
    Quartermaster
RM
    Radioman
RP
    Religious Program Specialist
SH
    Ship's Serviceman
SK
    Storekeeper
SM
    Signalman

ST
  Sonar Technician
STG
  Sonar Technician (surface)
STS
  Sonar Technician (submarine)
SW
  Steelworker
TD
  Tradesman
TM
  Torpedoman's Mate
UT
  Utilitiesman
WT
  Weapons Technician
YN
  Yeoman

COAST GUARD SPECIALTY RATINGS

AD
  Aviation Machinist's Mate
AE
  Aviation Electrician's Mate
AM
  Aviation Structural Mechanic
ASM
  Aviation Survivalman
AT
  Aviation Electronics Technician
BM
  Boatswain's Mate
DC
  Damage Controlman
DP
  Data Processing Technician
EM
  Electrician's Mate
ET
  Electronics Technician

MILITARY RANKS

FT
    Fire Control Technician
GM
    Gunner's Mate
HS
    Health Services Technician
IV
    Investigator
MK
    Machinery Technician
MST
    Marine Science Technician
PA
    Public Affairs Specialist
PS
    Port Securityman
QM
    Quartermaster
RD
    Radarman
RM
    Radioman
SK
    Storekeeper
SS
    Subsistence Specialist
TT
    Telephone Technician
YN
    Yeoman

CHAPTER 5

Missile and Rocket Designations

U.S. guided missiles are numbered in a single series; they are arranged in numerical sequence based on Table 4 below. The RUR-5A is a rocket-series weapon (listed at the end of Table 4).

All weapons listed have conventional warheads unless indicated otherwise.

MISSILE SERIES

AIM-7
   Sparrow (Raytheon and General Dynamics): Medium-range air-to-air missile. (USAF-USMC-USN)

RIM-7
   Sea Sparrow (Raytheon and General Dynamics): Adaptation of air-launched missile for shipboard use for anti-ship missile defense. Missile for NATO Sea Sparrow System (NSSM) and Basic Point Defense Missile System (BPDMS). (USN)

AIM-9
   Sidewinder (Ford Aerospace and Raytheon): Short-range air-to-air missile. (USAF-USMC-USN)

MIM-23
   Hawk (Raytheon): Medium-range, mobile surface-to-air missile. (USA-USMC)

LGM-30G
   Minuteman III (Boeing): Intercontinental Ballistic Missile (ICBM) fitted with multiple W78 nuclear warheads. (USAF)

Table 4. U.S. Missile and Rocket Designations

Explanation of symbols:

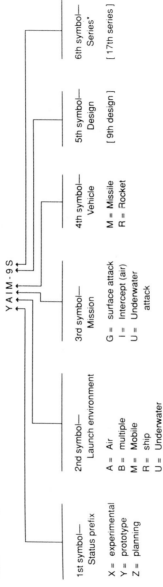

1st symbol—Status prefix	2nd symbol—Launch environment	3rd symbol—Mission	4th symbol—Vehicle	5th symbol—Design	6th symbol—Series*
X = experimental	A = Air	G = surface attack	M = Missile	[9th design]	[17th series]
Y = prototype	B = multiple	I = Intercept (air)	R = Rocket		
Z = planning	M = Mobile	U = Underwater attack			
	R = ship				
	U = Underwater				

*Note: Letters I and O are not used to avoid confusion with numerals.

SOURCE: Norman Polmar, *The Naval Institute Guide to the Ships and Aircraft of the U.S. Fleet*, 15th ed. (Annapolis: Naval Institute Press, 1993), 481.

MISSILE AND ROCKET DESIGNATIONS

MGM-31
Pershing II (Martin Marietta): Surface-to-surface nuclear battlefield missile; removed from service under the INF agreement; fitted with W50 nuclear warhead.

MIM-43
Redeye (General Dynamics/Pomona): Man-portable, short-range, surface-to-air missile; being replaced by the FIM-92. (USA-USMC)

MGM-51
Shillelagh (Ford Aerospace): Guided missile fired from 152-mm gun fitted in the Sheridan M551 armored vehicle. (USA)

MGM-52
Lance (LTV): Surface-to-surface battlefield attack missile designed for both nuclear and conventional warheads; being phased out of service. (USA)

AIM-54
Phoenix (Hughes): Long-range air-to-air missile carried by F-14 Tomcat. (USN)

AGM-62
Walleye II (Hughes/Martin Marietta): Air-to-ground tactical glide bomb. (USAF)

AGM-65
Maverick (Hughes): Air-to-ground, multi-role tactical missile. (USAF-USMC-USN).

RIM-66B
Standard-MR SM-1 (General Dynamics/Pomona): Ship-launched, medium-range, surface-to-air missile. (USN)

RIM-66C
Standard-MR SM-2 (General Dynamics/Pomona): Ship-launched, medium-range, surface-to-air missile. (USN)

RIM-67B
Standard-ER SM-2 (General Dynamics/Pomona): Extended range, improved version of ship-launched RIM-66C surface-to-air missile. (USN)

AGM-69A
SRAM (Boeing): Short-Range Attack Missile; air-to-ground strategic missile; W69 nuclear warhead. (USAF)

BGM-71
TOW (Hughes): Tube-launched, Optically tracked, Wire-guided anti-tank missile; can be launched from helicopters and various ground mountings. (USA-USMC)

MISSILE AND ROCKET DESIGNATIONS

MIM-72
Chaparral (Ford Aerospace): Short-range, surface-to-air missile. (USA)
FGM-77
Dragon (McDonnell Douglas): Man-portable assault missile for use against armor and bunkers. (USA-USMC)
AGM-84A
Harpoon (McDonnell Douglas): Air-launched anti-ship missile. (USN-USAF)
RGM-84A
Harpoon (McDonnell Douglas): Ship-launched anti-ship missile. (USN)
UGM-84A
Harpoon (McDonnell Douglas): Submarine-launched anti-ship missile. (USN)
AGM-84E
SLAM (McDonnell Douglas): Standoff Land-Attack Missile based on the Harpoon missile and with precision guidance for attacking ground targets. (USN)
AGM-86B
ALCM (Boeing): Air-Launched Cruise Missile fitted with W80 nuclear warhead. (USAF).
AGM-86C
ALCM (Boeing): Conventional version of AGM-86B. (USAF)
AGM-88A
HARM (Texas Instruments): High-speed Anti-Radiation Missile; advanced air-to-ground anti-radar weapon. (USAF-USMC-USN).
AIM-92
See FIM-92.
FIM-92
Stinger (General Dynamics/Pomona): Man-portable, short-range, surface-to-air missile. Fitted to helicopters as the AIM-92. (USA-USMC-USN).
UGM-96
Trident I C-5 (Lockheed): Submarine-Launched Ballistic Missile (SLBM); fitted with multiple W76 nuclear warheads. (USN)
MIM-104
Patriot (Raytheon): Medium-range, surface-to-air missile. (USA)
BGM-109
Tomahawk (General Dynamics/Convair and McDonnell Douglas): Ship- and submarine-launched, long-range cruise missile; the principal variants are Tomahawk Anti-Ship Missile (TASM) and

MISSILE AND ROCKET DESIGNATIONS

Tomahawk Land-Attack Missile (TLAM); the latter has several conventional warhead configurations and previously carried the W80 nuclear warhead in the TLAM-N configuration. The Ground-Launched Cruise Missile (GLCM) variant was discarded under the INF treaty. (USN)

AGM-114
Hellfire (Rockwell International and Martin Marietta): Helicopter-Launched Fire and Forget missile for use against tanks. (USA-USMC)

MIM-115
Roland (Hughes and Boeing): Short-range, surface-to-air missile. (USA)

RIM-116
RAM (General Dynamics/Pomona): Rolling Airframe Missile for shipboard use for anti-ship missile defense. (USN)

LGM-118
Peacekeeper (Martin Marietta): Intercontinental Ballistic Missile (ICBM); previously known as the MX; fitted with multiple W78 nuclear warheads. Being phased out. (USAF)

AGM-119B
Penguin (Norsk Forvarsteknologi A/S and Grumman): Helicopter-launched anti-ship missile. (USN)

AIM-120A
AMRAAM (Hughes and Raytheon): Advanced Medium-Range Air-to-Air Missile; replacement for AIM-7 Sparrow. (USAF-USMC-USN).

AGM-122
Sidearm (Motorola): Short-range, helicopter-launched anti-radar missile based on the AIM-9 Sidewinder. (USMC)

AGM-123
Skipper II (Aerojet/General Electric/Emerson): Laser-guided, rocket-propelled Mk 83 bomb made of off-the-shelf components. (USN)

UUM-125B
Sealance (Boeing): Proposed submarine-launched anti-submarine missile; development suspended. (USN)

AGM-129
ACM (Convair and McDonnell Douglas): Advanced Cruise Missile for strategic attack; fitted with nuclear warhead. (Planned USAF)

AGM-130
Mk 84 precision-guided bomb with "strap-on" rocket motor. (USAF)

MISSILE AND ROCKET DESIGNATIONS

AGM-131
SRAM II: Planned replacement for SRAM; cancelled in 1991.
AIM-132
ASRAAM: Advanced Short-Range Air-to-Air Missile planned to replace the AIM-9. (Planned USAF-USMC-USN)
UGM-133A
Trident II D-5 (Lockheed): Submarine-Launched Ballistic Missile (SLBM); fitted with multiple W88 nuclear warheads. (USN)
AGM-136
Tacit Rainbow (Northrop): Proposed "loitering" air-to-ground anti-radar missile; cancelled in 1991. (Planned USAF-USN)
AGM-137
Tri-Service Standoff Attack Missile (TSSAM) (Northrop-Boeing): Land-attack missile that can be launched from aircraft or ground launchers; possesses stealth characteristics. (Planned USA-USAF-USN)
RUM-139A
Vertical-Launch ASROC (Loral): Advanced ship-launched variant of the ASROC carrying the Mk 46 torpedo. (USN)
AGM-142
Have Nap/Pop Eye (Rafael): Israeli air-launched, stand-off land-attack missile. (USAF)

ROCKET SERIES

RUR-5A
ASROC (Honeywell): Anti-Submarine Rocket; fitted with conventional (Mk 46 torpedo). (USN)

CHAPTER 6

Ship Designations

ARMY

The Army uses a ship and craft designation series derived in part from the Navy's designation scheme:

BC
    barge, dry cargo (non–self-propelled)
BCDX
    barge, deck enclosure
BD
    floating crane
BDL
    beach discharge lighter
BG
    barge, liquid cargo (non–self-propelled)
BK
    barge, dry cargo (non–self-propelled)
BPL
    barge, pier, self-elevating
BR
    barge, refrigerated (non–self-propelled)
FMS
    floating marine repair shop (non–self-propelled)
FS
    freight and supply vessel (over 100 ft/30.48 m)

SHIP DESIGNATIONS

FSR
  freight and supply vessel, refrigerated
HLS
  heavy lift ship
J
  work boat (under 50 ft/15.24 m)
LARC
  lighter, amphibious, resupply, cargo
LCM
  landing craft, mechanized
LCU
  landing craft, utility
LCV
  landing craft, vehicle
LSV
  landing ship, vehicle
LT
  large tug (over 100 ft/30.48 m)
Q
  work boat (over 50 ft/15.24 m)
ST
  small tug (under 100ft/30.48 m)
T
  small freight and supply vessel (under 100ft/30.48 m)
Y
  liquid cargo vessel

COAST GUARD

The cutter designations currently in use for Coast Guard cutters and boats are:

WAGB
  icebreaker
WAGO
  oceanographic cutter
WHEC
  high endurance cutter (multi-mission; 30 to 45 days at sea without support)
WIX
  training cutter

SHIP DESIGNATIONS

WLB
   offshore buoy tender (full sea-keeping capability; medium endurance)
WLI
   inshore buoy tender (short endurance)
WLIC
   inland construction tender (short endurance)
WLM
   coastal buoy tender (medium endurance)
WLR
   river buoy tender (short endurance)
WLV
   light vessel
WMEC
   medium endurance cutter (multi-mission; 10 to 30 days at sea without support)
WPB
   patrol boat (multi-mission; 1 to 7 days at sea without support)
WSES
   surface effect ship

NATIONAL OCEANIC AND ATMOSPHERIC ADMINISTRATION

All NOAA ships are designated by a three-digit number proceeded by the letter R for Research and S for Survey, with the first digit indicating the Horsepower Tonnage (HPT) class. The HPT is the numerical sum of the vessel's shaft horsepower plus her gross tonnage.

NAVY

The following designations are from Secretary of the Navy Instruction (SECNAVINST) 5030.1, except for those designations indicated by an asterisk. The latter are in common usage, with some used by the U.S. Navy and U.S. intelligence community but are not official U.S. Navy ship designations.

Prefixes: Some basic ship designations are proceeded by a letter prefix used to indicate:
   F   being constructed for foreign government
   T-   assigned to Military Sealift Command
   W   Coast Guard vessel

SHIP DESIGNATIONS

The suffix *N* is used to denote nuclear-propelled ships. For service craft the suffix *N* indicates a non–self-propelled version of a similar self-propelled craft. Although the prefix letter *W* on the list indicates that Coast Guard classifications are included in the Navy list of classifications, in fact, they are different from Navy designations with few exceptions.

Note that parenthesis are not used for the suffix letter *N*, nor should hyphens be used between the designation and hull number.

ACS
    auxiliary crane ship
AD
    destroyer tender
AE
    ammunition ship
AF
    store ship
AFDB
    large auxiliary floating dry dock
AFDL
    small auxiliary floating dry dock
AFDM
    medium auxiliary floating dry dock
AFS
    combat store ship
AG
    miscellaneous
AGDS
    deep submergence support ship
AGF
    miscellaneous command ship
AGFF
    auxiliary general frigate
AGI*
    intelligence collection ship
AGM
    missile range instrumentation ship
AGOR
    oceanographic research ship
AGOS
    ocean surveillance ship

SHIP DESIGNATIONS

AGS
surveying ship
AGSS
auxiliary research submarine [not combat capable]
AH
hospital ship
AK
cargo ship
AKR
vehicle cargo ship
AO
oiler
AOE
fast combat support ship
AOG
gasoline tanker
AOR
replenishment oiler
AOT
transport oiler
AP
transport
APL
barracks craft
AR
repair ship
ARC
cable repairing ship
ARD
auxiliary repair dry dock
ARDM
medium auxiliary repair dry dock
ARL
repair ship, small
ARS
salvage ship
AS
submarine tender
ASR
submarine rescue ship
ATC
mini-armored troop carrier

SHIP DESIGNATIONS

ATF
    fleet ocean tug
ATS
    salvage and rescue ship
AVB
    aviation logistic support ship
AVM
    guided missile ship
AVT
    auxiliary aircraft landing training ship
BB
    battleship
CA
    gun cruiser
CG
    guided missile cruiser
CGN
    guided missile cruiser (nuclear-propelled)
COOP
    craft of opportunity
CRRC*
    combat rubber raiding craft
CT*
    COOP trainer
CV
    multi-purpose aircraft carrier
CVA
    attack aircraft carrier
CVN
    multi-purpose aircraft carrier (nuclear-propelled)
CVS
    ASW aircraft carrier
DD
    destroyer
DDG
    guided missile destroyer
DSRV
    deep submergence rescue vehicle
DSV
    deep submergence vehicle
FAC*
    fast attack craft

SHIP DESIGNATIONS

FF
   frigate
FFG
   guided missile frigate
FFL*
   light frigate
FFT
   frigate (reserve training)
FPB*
   fast patrol boat
FSS*
   fast sealift ship
HLPS*
   heavy-lift prepositioning ship
HSB*
   high-speed boat
IX
   unclassified miscellaneous (auxiliary)
LCAC
   landing craft, air cushion
LCC
   amphibious command ship
LCM
   landing craft, mechanized
LCPL
   landing craft, personnel, large
LCU
   landing craft, utility
LCVP
   landing craft, vehicle, personnel
LHA
   amphibious assault ship (general-purpose)
LHD
   amphibious assault ship (multi-purpose)
LKA
   amphibious cargo ship
LPA
   amphibious transport
LPD
   amphibious transport dock
LPH
   (1) amphibious assault ship (helicopter)
   (2) landing platform helicopter* [not used by U.S. Navy]

SHIP DESIGNATIONS

LSD
    dock landing ship
LSSC
    light SEAL support craft
LST
    tank landing ship
LWT
    amphibious warping tug
MCM
    mine countermeasures ship
MCMV*
    mine countermeasures vessel
MCT
    mine countermeasures craft, training (COOP)
MHC
    minehunter, coastal
MPS*
    maritime prepositioning ship
MSB
    minesweeping boat
MSL*
    minesweeping launch
MSO
    minesweeper—ocean
MSSC
    medium SEAL support craft
MTB*
    motor torpedo boat
NR
    submersible research vehicle (nuclear-propelled)
NTPS*
    near-term prepositioning ship
PB
    patrol boat
PBC*
    patrol boat, coastal
PBL*
    patrol boat, light
PBR
    river patrol craft
PC
    patrol, coastal

SHIP DESIGNATIONS

PCF
    patrol craft (fast)
PG
    patrol combatant
PHM
    patrol combatant missile (hydrofoil)
RHIB*
    rigid-hull inflatable boat
RIB*
    rubberized inflatable boat
RO/RO*
    roll-on/roll-off ship
SDV
    swimmer delivery vehicle
SLWT
    side-loading warping tug
SS
    submarine
SSAG*
    auxiliary submarine (combat capable)
SSB*
    ballistic missile submarine
SSBN
    ballistic missile submarine (nuclear-propelled)
SSG*
    guided missile submarine
SSGN
    guided missile submarine (nuclear-propelled)
SSN
    submarine (nuclear-propelled)
SSQ*
    communications submarine
SSQN*
    communications submarine (nuclear-propelled)
SWCL
    special warfare craft, light
SWCM
    special warfare craft, medium
YAG
    miscellaneous auxiliary service craft
YC
    open lighter

SHIP DESIGNATIONS

YCF
car float
YCV
aircraft transportation lighter
YD
floating crane
YDT
diving tender
YF
covered lighter
YFB
ferry boat or launch
YFD
yard floating dry dock
YFN
covered lighter
YFNB
large covered lighter
YFND
dry dock companion craft
YFNX
lighter (special purpose)
YFP
floating power barge
YFRN
refrigerated covered lighter
YFRT
range tender
YFU
harbor utility craft
YGN
garbage lighter
YLC
salvage lift craft
YM
dredge [self-propelled]
YMN
dredge [non–self-propelled]
YNG
gate craft
YO
fuel oil barge (self-propelled)

SHIP DESIGNATIONS

YOG
    gasoline barge (self-propelled)
YOGN
    gasoline barge
YON
    fuel oil barge
YOS
    oil storage barge
YP
    patrol craft, training
YPD
    floating pile driver
YR
    floating workshop
YRB
    repair and berthing barge
YRBM
    repair, berthing, and messing barge
YRDH
    floating dry dock workshop (hull)
YRDM
    floating dry dock workshop (machine)
YRR
    radiological repair barge
YRST
    salvage craft tender
YSR
    sludge removal barge
YTB
    large harbor tug
YTL
    small harbor tug
YTM
    medium harbor tug
YTT
    torpedo trials craft
YW
    water barge (self-propelled)
YWN
    water barge

About the Authors

Norman Polmar, an internationally known defense analyst and writer, is the author of more than a score of books on military, aviation, and naval subjects, including the Naval Institute's reference books *The Ships and Aircraft of the U.S. Fleet* and *Guide to the Soviet Navy.* He also writes a regular column for the Naval Institute *Proceedings.*

Mr. Polmar's interest in the education field includes his directing the Naval Institute's college intern program and serving on the

Norman Polmar (*center*) with his coauthors Mark Warren (*left*) and Eric Wertheim (*right*). *(Photo by Susan Thompson)*

advisory committee of the Elliott School of International Affairs of George Washington University in Washington, D.C.

Mark Warren and Eric Wertheim collaborated on this book while summer interns at the U.S. Naval Institute. Mr. Warren is a 1993 graduate of Stanford University, with a B.A. in political science. Mr. Wertheim, a junior majoring in history and political science at the American University in Washington, D.C., has also contributed to *Proceedings* magazine.

When this edition went to press, Mr. Warren had been accepted in the U.S. Navy's intelligence program and was attending aviation officer candidate school.

The **Naval Institute Press** is the book-publishing arm of the U.S. Naval Institute, a private, nonprofit society for sea service professionals and others who share an interest in naval and maritime affairs. Established in 1873 at the U.S. Naval Academy in Annapolis, Maryland, where its offices remain, today the Naval Institute has more than 100,000 members worldwide.

Members of the Naval Institute receive the influential monthly magazine *Proceedings* and discounts on fine nautical prints, ship and aircraft photos, and subscriptions to the bimonthly *Naval History* magazine. They also have access to the transcripts of the Institute's Oral History Program and get discounted admission to any of the Institute-sponsored seminars offered around the country.

The Naval Institute's book-publishing program, begun in 1898 with basic guides to naval practices, has broadened its scope in recent years to include books of more general interest. Now the Naval Institute Press publishes more than seventy titles each year, ranging from how-to books on boating and navigation to battle histories, biographies, ship and aircraft guides, and novels. Institute members receive discounts on the Press's nearly 400 books in print.

Full-time students are eligible for special half-price membership rates. Life memberships are also available.

For a free catalog describing Naval Institute Press books currently available, and for further information about U.S. Naval Institute membership, please write to:

Membership & Communications Department
U.S. Naval Institute
118 Maryland Avenue
Annapolis, Maryland 21402-5035

Or call, toll-free, (800) 233-USNI.